Ion-Beam-Based
Nanofabrication

MATERIALS RESEARCH SOCIETY
SYMPOSIUM PROCEEDINGS VOLUME 1020

Ion-Beam-Based Nanofabrication

Symposium held April 10–12, 2007, San Francisco, California, U.S.A.

EDITORS:

Daryush ILA
Alabama A&M University
Normal (Huntsville), Alabama, U.S.A.

John Baglin
IBM Almaden Research Center
San Jose, California, U.S.A.

Naoki Kishimoto
National Institute for Materials Science
Tsukuba, Ibaraki, Japan

Paul K. Chu
City University of Hong Kong
Kowloon, Hong Kong

Materials Research Society
Warrendale, Pennsylvania

Single article reprints from this publication are available through
University Microfilms Inc., 300 North Zeeb Road, Ann Arbor, Michigan 48106

CODEN: MRSPDH

Published by:

Materials Research Society
506 Keystone Drive
Warrendale, PA 15086
Telephone (724) 779-3003
Fax (724) 779-8313
Web site: http://www.mrs.org/

Manufactured in the United States of America

CONTENTS

*Invited Paper

PATTERNING, QUANTUM DOT SYNTHESIS, AND SELF ASSEMBLY

*Invited Paper

EXAMPLES: APPLICATIONS AND DEVICES

*Invited Paper

PREFACE

Symposium GG, "Ion-Beam-Based Nanofabrication," was held April 10–12 at the 2007 MRS Spring Meeting, in San Francisco, California.

Ion beam technologies now evidently offer a robust and highly versatile approach, both for commercial fabrication, and for nanometer-scale manipulation in research. The presentations during this two and a half day symposium emphasized applications of ion beams in nanoscale fabrication for custom tailoring of surface properties and features, and structures in 1-D, 2-D or 3-D, at resolution down to a few nanometers. Applications discussed included quantum dot arrays, nanopore membranes for DNA sequencing, bio-sensors and lab-on-chip devices, growth of nanowires, 3-D device integration, and high-density non-volatile memory. The presentations reported customized ion beam processing such as locally patterned surface functionalization to promote selective adhesion of cells in chosen arrays, or for biomedical implant materials; surface layer ultra-smoothing, by cluster ion bombardment; shallow implantation of dopants by ion clusters; induction of self-assembled surface ripples (a phenomenon of complex instability, whose full description is still eluding our best models); nanopore sculpting for electrophoretic DNA sequencing; growth of epitaxial nanowires on silicon; sculpting of intricate 3-D objects and arrays, including 'towers' with high aspect ratio; FIB-controlled growth of carbon nanotube devices; and ion implantation controlled development of silicon nanocrystals for photonic device applications.

The proceedings volume is divided according to the original sections of Symposium GG.

The organizers would like to take this opportunity to thank the sponsors, National Institute for Materials Science (NIMS), NASA, NEC, and MRS for providing financial support for this symposium.

Daryush ILA
John E.E. Baglin
Naoki Kishimoto
Paul K. Chu

July 2007

MATERIALS RESEARCH SOCIETY SYMPOSIUM PROCEEDINGS

Volume 989— Amorphous and Polycrystalline Thin-Film Silicon Science and Technology—2007, V. Chu, S. Miyazaki, A. Nathan, J. Yang, H.W. Zan, 2007, ISBN 978-1-55899-949-7

Volume 990— Materials, Processes, Integration and Reliability in Advanced Interconnects for Micro- and Nanoelectronics, Q. Lin, E.T. Ryan, W-L. Wu, D.Y. Yoon, 2007, ISBN 978-1-55899-950-3

Volume 991— Advances and Challenges in Chemical Mechanical Planarization, C. Borst, L. Economikos, A. Philipossian, G. Zwicker, 2007, ISBN 978-1-55899-951-0

Volume 992E— Deposition on Nonplanar Substrates, D. Josell, M. Brett, C. Witt, M. Ritala, 2007, ISBN 978-1-55899-952-7

Volume 993E— Pb-Free and RoHS-Compliant Materials and Processes for Microelectronics, E. Chason, 2007, ISBN 978-1-55899-953-4

Volume 994— Semiconductor Defect Engineering—Materials, Synthetic Structures and Devices II, S. Ashok, P. Kiesel, J. Chevallier, T. Ogino, 2007, ISBN 978-1-55899-954-1

Volume 995E— Extending Moore's Law with Advanced Channel Materials, S. Chakravarthi, R. Arghavani, G. Klimeck, 2007, ISBN 978-1-55899-955-8

Volume 996E— Characterization of Oxide/Semiconductor Interfaces for CMOS Technologies, Y. Chabal, A. Estève, N. Richard, G. Wilk, 2007, ISBN 978-1-55899-956-5

Volume 997— Materials and Processes for Nonvolatile Memories II, T. Li, Y. Fujisaki, J. Slaughter, D. Tsoukalas, 2007, ISBN 978-1-55899-957-2

Volume 998E —Nanoscale Magnetics and Device Applications, S.S. Xue, 2007, ISBN 978-1-55899-958-9

Volume 999E —Novel Semiconductor Materials for Room-Temperature Ferromagnetism, C.R. Abernathy, S. Bedair, P. Ruterana, R. Frazier, 2007, ISBN 978-1-55899-959-6

Volume 1000E—Functional Interfaces in Oxides, 2007, ISBN 978-1-55899-960-2

Volume 1001E—Progress in High-Temperature Superconductors, P. Barnes, D. Lee, C. Park, N. Amemiya, J. Reeves, 2007, ISBN 978-1-55899-961-9

Volume 1002E—Printing Methods for Electronics, Photonics, and Biomaterials, G. Gigli, 2007, ISBN 978-1-55899-962-6

Volume 1003E—Organic Thin-Film Electronics—Materials, Processes, and Applications, A.C. Arias, J.D. MacKenzie, A. Salleo, N. Tessler, 2007, ISBN 978-1-55899-963-3

Volume 1004E—Materials and Strategies for Lab-on-a-Chip—Biological Analysis, Microfactories, and Fluidic Assembly of Nanostructures, S. Grego, J.M. Ramsey, O. Velev, S. Verpoorte, 2007, ISBN 978-1-55899-964-0

Volume 1005E—Advances in Photo-Initiated Polymer Processes and Materials, A. Guymon, C. Hoyle, M. Shirai, E. Nelson, 2007, ISBN 978-1-55899-965-7

Volume 1006E—Transport Behavior in Heterogeneous Polymeric Materials and Composites, J. Grunlan, D. Bhattacharyya, E. Marand, O. Regev, A. Balazs, 2007, ISBN 978-1-55899-966-4

Snell, John L. 1776
Snoek, Johan 832
Snyder, Louis L. 365, 366, 367, 368, 369, 370
Sofer, Eugene 82
Solomian-Loc, Fanny 1024
Somerhausen, Anne 1510
Somerhausen, Christine 833
Šormová, Eva 1382
Sosnowski, Kiryl 1383
Soustelle, Jacques 834
Spanjaard, Barry A. 1384
Speer, Albert 371, 372, 1655
Spencer, Jack 1112
Sperber, Manes 1385
Spott, Frederic 373
Stachura, Peter D. 374, 375, 376, 377
Stadtler, Bea 1025
Stafford, David 836
Starkopf, Adam 1386
Staudinger, Hans 378a
Stein, George H. 378b
Steinbeck, John 1842
Steinberg, Lucien 1026, 1027
Steinberg, Milton 83
Steinbock, Johann 837
Steiner, Bedrich 1717b
Steiner, Erich G. 1028
Steiner, Jean-François 1029
Steinert, Marlis G. 379
Steinitz, Lucy Y. 1387
Stemper, Charles H. 84
Stenin, Afrikan A. 838
Stephenson, Jill 380
Stern, Ellen N. 1388
Stern, J.P. 381
Stevens, E.H. 1656
Stevenson, William 839, 1657
Stierlin, Helen 382
Stockwell, Rebecca 255
Storry, Richard 536
Strasser, Bernard P. 383
Strasser, Gregor 384
Strasser, Otto 385
Strauss, Herbert A. 1389
Strobinger, Rudolf 840
Strokach, Timofei 841
Strom, Margot S. 1390
Stroop, Jürgen 1391
Styron, William 1843

Suhl, Yuri 1030, 1031
Suzman, Arthur 1392
Svatá, Jarmila 1393
Swaab, Maurice 1777
Sweet-Escot, Bickham 842
Sydnor, Charles W. 386
Syrkin, Marie 1032
Szajkowski, Zosa 1394, 1718
Szmaglewska, Seweryna 1492

Tartakower, Aryel 991
Tauber, Kurt P. 387
Taylor, Alan J.P. 388
Taylor, John R. 1395
Taylor, Telford 389, 1658
Ten Boom, Corrie 940, 941
Tenenbaum, Joseph 1033, 1396
Teske, Hermann 843
Thalman, Rita 1397
Thomas, Catherine 390
Thomas, Gordon 1398
Thomas, Hugh 1659
Thomas, John (Jack) 844a
Thomas, John Oram 844b
Thompson, Dorothy 391
Thomsen, Erich 392
Thorne, Leon 1399
Thyssen, Fritz 393
Tillard, Paul 1278
Tillion, Germaine 942, 943
Tillon, C. 845
Tilton, Timothy A. 394
Timerman, Jacobo 85
Tokayer, Marvin 1400
Toland, John 395, 396
Tolischus, Otto D. 397
Toll, Nelly 1719
Tolstoy, Nikolai 398
Tomkiewicz, Mina 1844
Treece, Patricia 846
Trenowden, Ian 847
Trepman, Paul 1401
Trepper, Leopold 848
Trevor-Roper, H.R. 399, 400, 401
Trivanovitch, Vaso 1660
Troper, Harold 1892
Trouillé, P. 849
Trunk, Isaiah 1402, 1403, 1404
Tsatsos, Jeanne 850
Tsion, Daniel 1405

MATERIALS RESEARCH SOCIETY SYMPOSIUM PROCEEDINGS

Prior Materials Research Society Symposium Proceedings available by contacting Materials Research Society

Ion Beam Nanofab:
Tools, Techniques, and Applications

Mater. Res. Soc. Symp. Proc. Vol. 1020 © 2007 Materials Research Society   1020-GG01-02

Cluster Ion Beam Process for Nanofabrication

Isao Yamada, and Noriaki Toyoda
Graduate School of Engineering, University of Hyogo, 2167 Shosha, Himeji, 671-2280, Japan

Abstract. This paper reviews gas cluster ion beam (GCIB) technology, including the generation of cluster beams, fundamental characteristics of cluster ion to solid surface interactions, emerging industrial applications, and identification of some of the significant events which occurred as the technology has evolved into what it is today.   More than 20 years have passed since the author (I.Y) first began to explore feasibility of processing by gas cluster ion beams at the Ion Beam Engineering Experimental Laboratory of Kyoto University.   Processes employing ions of gaseous material clusters comprised of a few hundred to many thousand atoms are now being developed into a new field of ion beam technology.   Cluster-surface collisions produce important non-linear effects which are being applied to shallow junction formation, to etching and smoothing of semiconductors, metals, and dielectrics, to assisted formation of thin films with nano-scale accuracy, and to other surface modification applications.

HISTORICAL MILESTONES IN GCIB TECHNOLOGY

In 1950, Becker et al first studied cluster formation for thermonuclear fuel applications using gaseous materials passed through supersonic nozzles wchichi were cooled by liquid nitrogen and helium shrouds [1]. The supersonic expansion approach was successful in producing cryogenic beams containing large numbers of clusters. This original work opened the way to employ gas clusters for materials processing.

During the late 1970's and 1980's, an ionized cluster beam (ICB) technique which employed metal vapor clusters from heated Knudsen cells for thin film formation was studied at Kyoto University and elsewhere.   Kyoto University investigations of metal vapor clusters ended when collaborative work with W.L.Brown at Bell Laboratories showed the cluster ion intensities within the metal vapor streams to be too low for most practical purposes [2,3].   Subsequent work at the Kyoto University Ion Beam Engineering Experimental Laboratory then focused upon cluster beam formation employing gas expansion through simple supersonic nozzles.

Initial research on gas cluster beam formation showed that supersonic nozzles having converging-diverging shapes operating at room temperature could produce intense beams of gas clusters.   This then led to research and development of gas cluster ion beam (GCIB) techniques [4] and to investigations of new ion-solid interactions produced by gas cluster ion impacts. These studies demonstrated that GCIB produces unique ion/solid interactions and offers new atomic and molecular ion beam process opportunities in areas of implantation, sputtering, and ion beam assisted deposition.   Most of the original technical results through to the year 2000 have been summarized in a monograph. [5].

Over the first 10 years of GCIB studies, low energy surface interaction effects, lateral sputtering phenomena and high chemical reaction effects were observed experimentally and were explained by means of molecular dynamics (MD) modeling. Japanese government funding through JST (Japan Science and Technology), MEXT (Agency, Ministry of Education), METI (Ministry of Economy, Trade and Industry) and others provided long term support for the research at Kyoto University. Difficulty in developing GCIB equipment within Japan resulted in development of commercial GCIB equipment by Epion Corporation in the U.S. beginning in 1995 [6].

In 2000, a four year R&D project for development of GCIB industrial technology began in Japan under funding from the New Energy and Industrial Technology Development Organization (NEDO). This project involved subjects in areas of semiconductor surface processing, high accuracy surface processing and high-quality film formation. The project was supported by the formation of a new Collaborative Research Center of Cluster Ion Beam Technology at Kyoto University and University of Hyogo.

In 2002, another major GCIB project which emphasized nano-technology applications was started under a contract from the Ministry of Economy and Technology for Industry (METI). This METI project currently involves development related to size-selected cluster ion beam equipment and processes, and development of GCIB processes for very high rate etching and for zero damage etching of magnetic materials and compound semiconductor materials.

Figure 1 shows historical milestones of GCIB equipment and process development.

GAS CLUSTER FORMATION AND GCIB EQUIPMENT

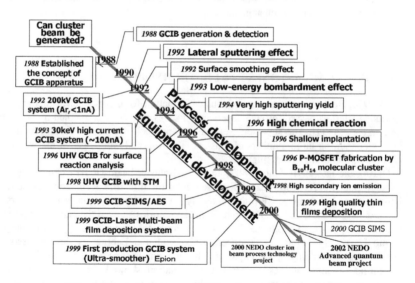

Figure 1. Historical milestones of GCIB equipment and process development.

GCIB processing of materials is based on the use of electrically charged cluster ions consisting of a few hundred to a few thousand atoms or molecules of gaseous materials. A beam of neutral clusters is first formed from individual gas atoms by expansion of the gas through a nozzle at room temperature into vacuum. The clusters are subsequently ionized and accelerated.

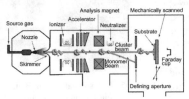

Figure 2. Typical configuration of GCIB equipment

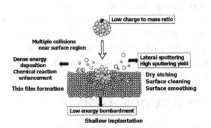

Figure 3. GCIB-solid surface interactions and their applications.

The typical configuration of GCIB equipment is as shown in Figure 2. A small aperture, or skimmer, transmits the primary jet core of gas clusters emerging from the expansion nozzle. Forward-directed neutral clusters are then ionized by impact of electrons accelerated from a filament so as to form positive ion gas clusters with nominally one charge per cluster. The ionized clusters are extracted and accelerated through typical potentials of between and 2 and 30 kV using a series of electrodes. Electrostatic lenses are utilized to focus the cluster ions, and monomers are filtered out by means of a strong transverse magnetic field. Usually the cluster ion beam is kept stationary and material to be processed is scanned mechanically through the beam so as to obtain uniform and complete coverage. The cluster ion fluence is measured by means of a Faraday cup.

When an energetic cluster ion impacts upon a surface, it interacts nearly simultaneously with many target atoms and deposits high energy density into a very small volume of the target material. The concurrent energetic interactions between many atoms comprising the cluster and many atoms of the target result in highly non-linear implantation and sputtering effects [7]. These effects, which are fundamentally different from those associated with the more simple binary collisions occurring during monomer ion impacts, include low energy bombardment phenomena, lateral sputtering effects and high chemical reaction effects. Figure 3 summarizes GCIB-solid surface interactions and their applications.

In early 1988, cluster formation was confirmed by electron diffraction. A strong Debye-Scherrer ring, as shown in Figure 4(a), was observed when electrons were diffracted through the beam emerging from a gas nozzle. Time of flight (TOF) measurements were subsequently used to show that the clusters contained several hundred to many thousand atoms [8]. The clusters were believed to be held together by van der Waals forces.

Figure 4(b) shows typical size distributions found for Ar gas clusters. Clusters of the sizes which are normally produced by room temperature nozzles were found to be particularly useful for materials processing. More recent

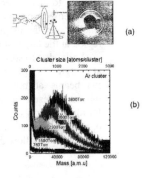

Figure 4. Cluster detection by e-diffraction (a) and TOF (b).

experiments conducted with size-selected cluster beams have confirmed the fortuitous nature of the cluster size distributions which are typically produced [9]. It has been found that clusters can be formed from nearly all gases and gas mixtures, including rare gases such as Ar and Xe, most diatomic gases such as O_2 and N_2, and molecular compound gases such as B_2H_6, BF_3, CH_4, NF_3, SF_6, etc.

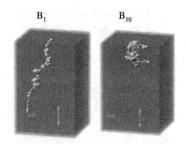

Figure 5. MD simulations of B monomer and B_{10} cluster ion implantations into Si at 5 keV.

One of the advantages associated with the cluster ion is a very low charge to mass ratio. Cluster ions containing up to several thousands of atoms typically become only singly or doubly ionized. Consequently, a cluster ion beam at any given current density can transport up to thousands of times more atoms than a monomer ion beam at the same current density. For example, a 1µA beam of cluster ions with average size of 1000 atoms per cluster can transport the same number of atoms as a 1mA monomer ion beam. These characteristics make it possible to use GCIB for very low energy ion beam processes which are normally difficult by traditional ion beam technology. Available GCIB equipment is now able to produce cluster ion beam currents of hundreds of microamperes or more from gases such as Ar [6].

LOW ENERGY EFFECTS AND SHALLOW JUNCTION FORMATION

From the beginning of cluster ion beam investigations, it was expected that clusters should produce low energy bombardment effects since the kinetic energy of each atom in a cluster ion is roughly equal to the total energy of the cluster divided by the number of atoms contained in the cluster. As an example, within a 20 keV cluster ion consisting of 2000 atoms, each of the individual atoms has energy of only 10 eV. Low energy effects were predicted by MD simulations and were confirmed by experiments, for example by comparing B monomer ion and B cluster ion implantation [10]. The results showed low-energy individual atomic interactions even when the total energy of the clusters was high. Figure 5 shows MD simulations of B monomer and cluster ions into Si illustrating the low energy effect of cluster ion bombardment relative to monomer ion bombardment. Important differences in range and density of the displacements produced are apparent. In the cluster ion case, the penetration range is extremely shallow and the displacements that are produced remain tightly concentrated within the impact region at the target surface.

While, due to space charge effects, it is exceptionally difficult to transport monomer ion beams at energies as low as 10 eV, equivalently low energy ion beams can be realized by using cluster ion beams at high acceleration voltages. The standard configuration of GCIB equipment inherently results in highly directional parallel cluster ion beams which are extremely well suited for ultra shallow doping applications and are known to produce thin film transistors which exhibit operational characteristics superior to those which result when conventional ion implantation is used for the shallow doping [11].

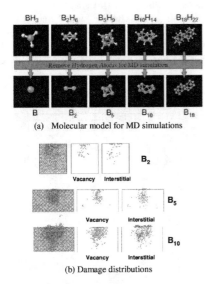

(a) Molecular model for MD simulations

(b) Damage distributions

Figure 6. MD simulation models of B clusters (a) and typical implant damage distributions for B_2, B_5 and B_{10} ions (acceleration energy: 500eV/atom)

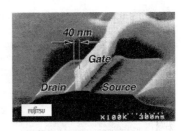

Figure 7. First p-MOSFET by $B_{10}H_{14}$ implantation

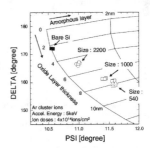

Figure 8. Cluster size dependence of Ψ and Δ after 5keV Ar-GCIB irradiation with ion dose of 4×10^{14}ions/cm^2.

POLYATOMIC CLUSTER IMPLANTATION

The concept of performing polyatomic cluster implantation into Si by using decaborane ($B_{10}H_{14}$) molecules was first investigated in 1996 by Kyoto University researchers working together with researchers at Fujitsu [12]. Effects upon range and damage distributions resulting from implantation of different sizes of B polyatomic clusters have been studied by MD simulations employing B_1, B_2, B_5, B_{10}, B_{18} molecular models shown in Figure 6(a) [13]. The results of these simulations indicate that cluster-like bombardment phenomena, the nonlinear effects which are typical of cluster impact, begin to be observed with clusters containing at least five or more atoms. Figure 6(b) shows the implant damage distributions resulting from the model clusters of 2 atoms, 5 atoms and 10 atoms. From MD simulations of bombardment by even larger clusters, it has been shown that the displacement damage increases with increasing cluster size and very clusters, as in the case of GCIB, cause complete self amorphization [5].

P-MOSFETs with 40nm gates, as in the device shown in Figure 7, were successfully fabricated by Fujitsu in 1996 using $B_{10}H_{14}$ implantation for ultra shallow junction and source/drain formation [12]. Decaborane implantation at 30 keV to a dose of 1×10^{13} ions/cm^2 followed by annealing at 1000°C for 10 seconds was reported to have resulted in a junction depth of 20 nm. For source/drain extensions, $B_{10}H_{14}$ ion implantation at 2 keV was carried out to a dose of 1×10^{12} ions/cm^2 followed by annealing at 900°C for 10 seconds.

Under a 2002 contract from JST (Japan Science and Technology Agency) Nissin Ion Corporation has successfully developed equipment for $B_{10}H_{14}$ ion implantation. A beam current of 3 mA at an acceleration energy of 3 keV has been achieved [14]. Using this equipment, Nissin and Fujitsu have worked to advance the decaborane process technology. Recently they have reported devices with significantly reduced threshold voltage deviation and higher forward currents than similar devices made by conventional B implantation [11]. Other polyatomic cluster implantation, such as that using $B_{18}H_{22}$, has also been developed because of the decaborane success [15].

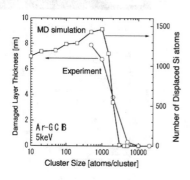

Figure 9. Cluster size dependence of number of displaced Si atoms by MD simulations and damage layer thickness

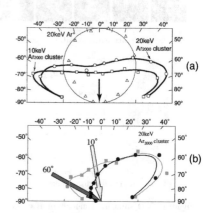

Figure 10. Angular distribution of sputtered atoms by Ar monomer and Ar cluster ions.(a) normal incidence (b) oblique incidence.

CLUSTER SIZE EFFECTS AND LOW DAMAGE PROCESSES

The influence of cluster size upon surface damage production has been studied experimentally using GCIB apparatus which incorporates a strong permanent sector magnet for cluster size

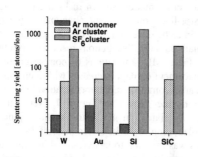

Figure 11. Reactive and physical sputtering for various materials at 20keV acceleration energy.

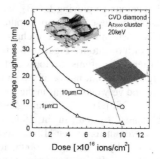

Figure 12. Ion dose dependence of the average roughness of a CVD diamond surface bombarded with a 20keV Ar cluster ion beam at normal incidence

selection [9]. Displacement damage within the Si surfaces has been evaluated by ellipsometry using a two-layer analysis model which assumes that oxide and amorphous layers are formed by the GCIB bombardment. From ellipsometry measurements, the intensity ratio Ψ and phase difference Δ of p and s waves can be determined. An increase in amorphous layer thickness is indicated by an increase of Ψ and an increase in the oxide layer thickness is indicated by a decrease of Δ. From the Ψ and Δ behaviors, estimates can be made of the damage formation due to Ar-GCIB bombardment.

Studies have been made of Si surfaces bombarded by size-selected 5 keV Ar cluster ion beams to dose levels of $4x10^{14}$ ions/cm^2. Mean cluster sizes utilized were 540, 1000 and 2200 atoms per cluster, resulting in average energies of 9.2, 5.0 and 2.3 eV per atom respectively. At the fixed 5 keV beam energy, both the oxide thickness and the amorphous layer thickness were found to decrease monotonically with increasing cluster size, ie., with decreasing energy of the cluster atoms. Figure 8 shows Ψ and Δ plots after Ar-GCIB irradiation of the Si.

Figure 9 shows plots of estimated damaged layer thickness and total displacements versus cluster size as determined experimentally and also by MD simulations [9]. From the MD simulations, the number of Si atoms displaced from their lattice cites increased with Ar cluster size from 10 to 1000, showed a peak at around 1000 atoms/cluster and then rapidly decreased with further increase of cluster size. At cluster sizes above 4000 atoms per cluster, MD simulations showed almost no displaced atoms after Ar-GCIB irradiation. The experimental results have shown almost the same trend as the MD simulations. These results suggest that shallow implantation and doping can be possible by using large cluster ions. From MD simulations, it is predicted that there is no damage formation in Si when the energy per atom in Ar cluster ion is below 1eV even though large cluster ions having such conditions can still deposit approximately 30% of their acceleration energy into a target Si substrate.

LATERAL SPUTTERING AND ATOMIC LEVEL SMOOTHING

An important characteristic of large gas-cluster ion bombardment is an effect known as lateral sputtering. Angular distributions of surface atoms ejected by cluster ions are considerably different from the distributions produced by monomer ions. Figure 10 shows experimentally measured angular distributions of sputtered atoms by Ar_{2000} cluster ions at (a) normal incidence and (b) oblique incidence. The angular distribution produced by monomer ions, which indicates the usual cosine distribution, is also shown. The angular distribution of sputtered atoms produced by the Ar cluster ions illustrates the lateral ejection. [8].

Sputtering yields due to cluster ions are very high relative to those associated with monomer ions at similar energy. Very high sputtering yields on metal, semiconductor and insulator surfaces due to bombardment with cluster ions have been observed experimentally. Experimentally measured sputtering yields of various materials due to 20 keV Ar (physical) and SF$_6$ (chemically reactive) cluster ions and monomer ions are shown in Figure 11.

Lateral sputtering produces surface smoothing behavior which does not occur with monomer ions. Smoothing of surfaces to atomic levels has been the first production use for cluster ion beam processing. Figure 12 shows typical ion dose dependence of the average roughness of a CVD diamond surface bombarded with a 20 keV Ar cluster ion beam at the normal incidence. The average roughness decreased monotonically with increasing ion dose from the initial value of 26 nm to a value of 1.3 nm after an ion dose of $8x10^{16}$ ions/cm^2. In the case of monomer ion irradiation at normal incidence, surface roughness typically increases with ion dose due to erosion or bubble formation inside the target.

GCIB APPLICATIONS IN INDUSTRY

Epion Corporation in the U.S. has developed GCIB equipment for industrial applications under a license from Japan Science and Technology Agency (JST). As is shown in Figure 13, concurrent with the development of increasingly capable commercial GCIB equipment has been the development of applications for GCIB process technology in the manufacturing of semiconductor devices and other advanced technical devices [16].

GCIB applications in semiconductors

As is suggested in Figure 14, a number of candidate applications are being developed for GCIB in the manufacturing of coming generations of semiconductor devices.   These applications include ultra shallow junction doping, SiGe alloy formation, film deposition, silicon-on-insulator thinning and uniformity correction, etching of dielectrics, ashing of photoresist, and surface modification of metals and dielectrics for integration of improved Cu interconnects.

GCIB offers excellent characteristics for producing ultra shallow junctions with pre-activation depths ranging from less than 10 nm to a maximum of approximately 30 nm.   The mechanism of doping by GCIB, which is referred to as "infusion", depends upon the intense localized temperature/pressure transient which is created at the point of cluster impact.   During the moment of impact, molecular gases such as B_2H_6 contained within the cluster undergo dissociation and solid species such as B which they contain then undergo mixing into the target

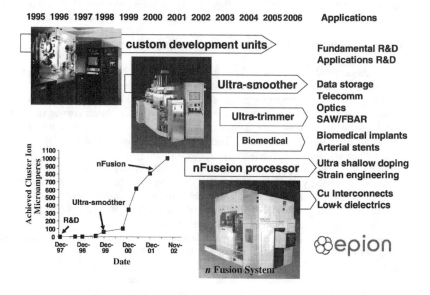

Figure 13. Industrial GCIB application areas

within the region of the thermal transient while gaseous atoms escape from the surface. Unlike ion implantation, the junction depths resulting by GCIB infusion depend upon the 1/3rd power of the cluster ion energy. Although the cluster beams used for ultra shallow doping are formed using dilute concentrations of dopant gas within an inert carrier gas host, for example 1% B_2H_6 in 99% Ar, the GCIB infusion processes are very efficient and throughputs are very high.

GCIB processes employing reactive gases such as halogen compounds can be used to produce very uniform and reproducible chemical etching. An optional mode of operation of GCIB equipment allows very precise but intentionally nonuniform "corrective etching" processes to be performed. One example application which has been demonstrated for this controlled nonuniniform processing capability of GCIB is in corrective etching and smoothing to reduce the thickness and improve the uniformity of active silicon layers on silicon-on-insulator (SOI) materials. As an example, a 20 keV 5 % CF_4 with 95 % O_2 halogen-GCIB process and a 3 keV O_2-GCIB process were used to provide rapid corrective etching and smoothing of a 150 mm SOI wafer. The 145 nm thick Si surface layer with standard deviation of 0.85 nm was reduced to 50 nm with standard deviation of 0.4 nm. In order to fully realize the benefits of SOI substrates for fully depleted MOS devices, Si layers only 20 nm or less thick with thickness nonuniformity of <5 % are desired. A 50 nm thick Si layer having greater than 3 nm nonuniformity over a 300mm SOI wafer can typically be reduced to 20 nm thickness with less than 0.4 nm of nonuniformity.

GCIB densification of porous materials is a new ion beam process which would not be expected to be possible by monomer or molecular ion beams. Argon GCIB processing of porous low-k dielectric materials was investigated on blanket spin-on methyl-silsesquioxane (p-MSQ) films of k~2.2 on 200 mm Si wafers by Epion Corporation and International Sematech. The results showed that a GCIB-densified surface could be formed without alteration of the dielectric

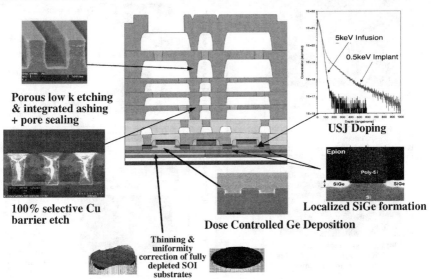

Figure 14. GCIB applications in semiconductors

material composition. The densified surface produced on the low-k dielectric was able to prevent penetration into the dielectric by Ti from a subsequently deposited PECVD TiSiN film, whereas low-k materials without the surface densification showed Ti penetration under the same conditions.

The highly reactive yet substrate sensitive properties of GCIB have allowed it to be very effective for ashing with high rates of polymer removal without any etching or degradation of other exposed materials such as porous low-k dielectrics. When reactive gases such as O_2, N_2, or C-containing molecules are included in the beams, surface reactions can take place to form thin films, and these films can be used in devices. Chemical GCIB can also be used for cleaning surfaces such as Cu or for etching of thin films in normally hard to reach surfaces within a semiconductor device, such as at the bottom of a trench or via. A GCIB high energy directed chemical beam can perform tasks that are difficult or impossible by other technologies such as plasma processing.

GCIB applications to non-semiconductor fields

Currently, industrial applications in several non-semiconductor fields are being developed by a number of Japanese companies under the nanotechnology program called "Advanced Nano-Fabrication Process Technology Using Quantum Beams" of NEDO /METI (New Energy and Industrial Technology Development Organization /the Ministry of economy and Technology Industry). Figure 15 summarizes the project.

GCIB surface smoothing processes are being applied for surfaces of magnetic materials used

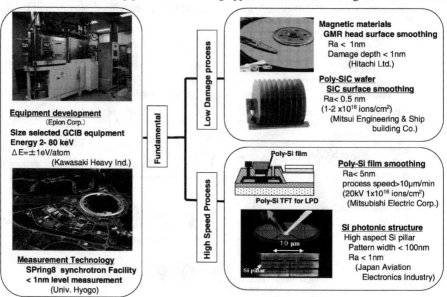

Figure 15. Summary of NEDO/MITI project.

for HDD sensor heads (Hitachi, Ltd.), for surfaces of polycrystalline SiC wafers which are used as monitor wafers for CVD processes (Mitsui Engineering & Shipbuilding Co., Ltd), for laser annealed poly-Si surfaces which are used for flat panel displays (Mitsubishi Electric Corp.) and for Si nano-structure surfaces for photonics (Japan Aviation Electronics Industry, Ltd.)

One of the more remarkable applications under development involves surface smoothing of the side walls of high aspect ratio Si pillar structures for photonic devices. The successive etching and deposition by inductively coupled plasma – reactive ion etching (ICP-RIE) used to fabricate the tall pillar structures results in side walls which are excessively rough. A GCIB smoothing process has been used to smooth the side wall surfaces to an Ra value of 0.1 nm.

SUMMARY

The history and present status of research and development in the field of gas cluster ion beam processing has been reviewed. Non-linear and non-equilibrium effects due to bombardment of large cluster ions are now attracting attention as new technology in the area of ion beam processing. GCIB processing is an advanced approach which will contribute to further progress in the ion beam technology field. Some industrial applications of GCIB which are being developed have been discussed.

ACKNOWLEDGEMENTS

The authors wish to thank to A.Kirkpatrick of Epion Corporation for long collaboration in the development of commercial GCIB equipment and applications. This work is partially supported by New Energy and Industrial Technology Development Organization (NEDO).

REFERENCES

* Emeritus Professor, Kyoto University

1. E.W.Becker 'On the history of cluster beams', in F.Trager and G. zu Putlitz, eds, Proc. Internat. Symp. On Metal Clusters- 1986, Springer Verlag, Berlin (1986) p. 1
2. R.L.McEachern, W.L.Brown, M.F.Jarrold, M.Sosnowski, G.H.Takaoka, H.Usui and I.Yamada, J. Vac. Sci. Technol. A, 9 (1991) 3105.
3. W.L.Brown, M.F.Jarrold, R.L.McEachern, M.Sosnowski, G.H.Takaoka, H.Usui and I.Yamada, Nucl. Instr. and Meth. B, 59/60 (1991) 182.
4. I.Yamada, Radiation Effects and Defects in Solids, 124 (1992) 69.
5. I.Yamada, J.Matsuo, N.Toyoda, and A.Kirkpatrick, Materials Science and Engineering R34, (2001) 231.
6. A.Kirkpatrick Extended abstracts, Workshop on cluster ion beam process technology, Kyoto International Community House, Kyoto October 12-13, 2000, Osaka Science & Technology Center, (2000) p.17.
7. J.Matsuo, Kyoto University PhD thesis (2000) and T.Seki Kyoto University PhD thesis (2000)
8. N.Toyoda, Kyoto University PhD thesis (1999).
9. N.Toyoda, S.Houzumi, T.Aoki and I.Yamada, Mat. Res. Soc. Symp. Proc. 792, (2004) p.623.
10. T.Aoki, Kyoto University PhD thesis (2000)

11. T.Aoyama, S.Umisedo, N.Hamamoto, T.Nagayama, M.Tanjyo, Extended Abstracts of the Sixth International Workshop on Junction Technology, Shanghai, May 15-16, 2006 , IEEE Press, (2006)p.88.
12. K.Goto, J.Matsuo, T.Sugii, H.Minakata, I.Yamada, IEDM Tech Dig. 1996, IEEE (1996) p. 435. and K.Goto, J.Matsuo, Y.Tada, T,Tanaka, Y.Momiyama, T.Sugii, and I.Yamada, IEDM Tech. Dig., IEEE (1997), p. 471.
13. T.Aoki Private communication
14. Japan Science and Technology Agency (JST)   Project completed on 2005-06-18.
15. D.Jacobson, Extended abstracts of the fifth international workshop on junction technology, June 7-8 2005, Osaka Japan, Japan Society of Applied Physics/Silicon Technology division. IEEE, (2005) p23
16. A.Kirkpatrick Extended abstracts, 6th workshop on cluster ion beam and advanced quantum beam process technology, KKR Hotel Tokyo, September 26-27, 2005, Osaka Science & Technology Center, (2005) p.26.

Mater. Res. Soc. Symp. Proc. Vol. 1020 © 2007 Materials Research Society   1020-GG01-05

Recent Advances in FIB Technology for Nano-prototyping and Nano-characterisation

Debbie J Stokes[1], Laurent Roussel[1], Oliver Wilhelmi[1], Lucille A Giannuzzi[2], and Dominique HW Hubert[1]
[1]FEI Company, Eindhoven, 5600, Netherlands
[2]FEI Company, Hillsboro, OR, 97124

ABSTRACT

Combined focused ion beam (FIB) and scanning electron microscopy (SEM) methods are becoming increasingly important for nano-materials applications as we continue to develop ways to exploit the complex interplay between primary ion and electron beams and the substrate, in addition to the various subtle relationships with gaseous intermediaries.

We demonstrate some of the recent progress that has been made concerning FIB SEM processing of both conductive and insulating materials for state-of-the-art nanofabrication and prototyping and superior-quality specimen preparation for ultra-high resolution scanning transmission electron microscopy (STEM) and transmission electron microscopy (TEM) imaging and related *in situ* nanoanalysis techniques.

INTRODUCTION

State-of-the-art focused ion beam technology combined with high-performance scanning electron microscopy is making a big impact, particularly with the ability to use either focused ions or electrons to perform a number of different tasks at the nano-scale. A typical arrangement of beams is shown in figure 1 below.

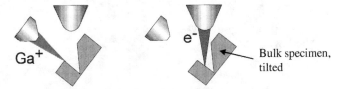

Figure 1. Schematic diagram to show the relative geometries of the specimen and ion and electron beams in a FIB SEM system. Gallium is commonly used as a FIB source.

Achieving the highest standards requires an understanding of the physics and chemistry of the system as a whole, which contains ions and electrons of various energies and origins, substrates with a range of electrical and mechanical properties, and reactive gases capable of specific effects on *in situ* chemical vapor deposition, sputtering and redeposition. We have built up a detailed knowledge of these complex parameters and, in this paper, we discuss three aspects of FIB processing (1) nano-fabrication and prototyping (2) milling of non-conductive materials and (3) preparation of specimens for use with other techniques.

DISCUSSION

Nanofabrication and prototyping

The milling of patterns in any kind of material and the precise deposition of various metals or non-metallic materials in a single instrument are recognized as novel means for truly rapid prototyping. An on-board digital pattern generator can be used for steering the ion or electron beam according to a given design. At present, pattern generators for electron beam lithography (EBL) are being used together with their dedicated software on FIB SEM instruments. However, electron-beam lithography exposure strategies do not take into account the fundamental differences in beam-substrate interactions.

In EBL, electron beam exposure creates a latent image in a resist film, which is developed once the pattern is completed and the substrate has been removed from the instrument. The resulting structures are a convolution of the dose accumulated in the resist and the contrast function of the resist-developer system.

By contrast, milling and deposition processes in a FIB SEM instrument arise from direct patterning techniques.. Consequently, milling rates and the uniformity of patterns depend on the exposure strategy, for example: dwell time, refresh time, single-pass or multi-pass execution of individual pattern elements, the definition of leading edges, consideration of redeposition in the milling order and a material-dependent selection of the pitch (i.e., overlap) are all important aspects for successful prototyping. The erosion of fiducials in overlay exposures, the effects of breaking down individual pattern elements, time considerations and beam blanking are further factors to be taken into account.

Proper evaluation of the relevant factors has a dramatic effect on the precision of FIB-patterning and chemical vapor deposition CVD, and is essential for high-quality fabrication of structures such as Fresnel zone plate lenses [1], photonic arrays [2] and deposition of 3-dimensional structures [3]. As an example of the importance of different milling strategies, figures 2(a)–(c) help to illustrate the results of three different approaches.

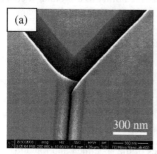 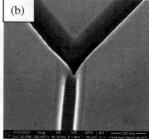 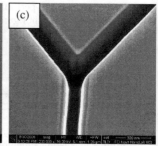

Figure 2. FIB-milled structures in silicon using three different milling strategies: (a) single pass milling (b) multi-pass, serial milling and (c) multi-pass, parallel milling.

In figure 2(a) an EBL approach has been used, in which FIB milling was performed in a single pass of the beam (dwell time per pixel = 882 μs). The result exhibits pronounced milling artifacts which reflect the direction of the sweeps and the leading edge. Redeposition on the sidewalls leads to a v-shaped line profile and artifacts at the junction. Figure 2(b) shows the same pattern milled using short pixel dwell times (1 μs), to avoid the formation of strong topography in one pass, and multi-pass milling to achieve the required depth (total time = 882 μs). The features of the pattern have been milled in a serial manner (one after another), and hence some redeposition occurs, leading to artifacts at the junction. Figure 2(c) shows the result when the pattern is executed using short dwell times and multi-pass milling, as in figure 2(b), but in a parallel manner. Milling of the entire pattern until completion prevents the build-up of redeposition and leads to a high-quality result, and a trench pattern that could be directly used in a fluidic device, for instance.

Progress in FIB milling of electrically insulating materials

There are many potential applications of FIB SEM involving soft and/or electrically insulating materials such as polymers, ceramics, glasses and biological specimens. However, these are somewhat demanding due, to a large extent, to the accumulation of positive charge during ion beam irradiation and the further formation of a positive surface potential as a result of secondary electron emission. In order to truly extend our capabilities in this regard, we must address the issue, since the resulting image drift can have significant consequences upon the accuracy and quality of FIB milling, imaging and CVD. Whilst the application of an electrically conductive coating to the specimen surface can be adequate to overcome charging problems associated with the primary ion beam, it is not always appropriate or practical to apply such a coating, and so other methods of charge control are needed. For example, on a single beam FIB instrument, an ancillary electron 'flood gun' is typically used to deliver low energy electrons to the specimen surface, thereby helping to maintain a charge balance. The flood gun method is also used for secondary ion mass spectroscopy (SIMS), x-ray photoemission spectroscopy (XPS), Auger electron spectroscopy (AES) and in most FIB SEM systems operating with a field emission electron source.

We have developed a method for suppressing ion beam drift using a defocused, low energy primary electron beam, and have derived an analytical method to correlate the ion and electron beam energies and currents with other parameters required for electrically stabilising these challenging materials [4]. Thus we are able to apply this knowledge to create high resolution microstructures such as nano- and micro-fluidic channels in electrically insulating substrates, as can be seen in figure 3. In addition, such control enables us to successfully perform unattended or automated FIB-based operations on insulating materials. The quartz specimen shown in figure 3 was made using the same multi-pass, parallel milling strategy discussed for silicon in the previous section. The side walls are parallel, the channels are clean and the top surface is clear of re-deposited material, suggesting that the *en face* fusing or adhesion of two similar specimens would result in a functional, optically transparent, nanofluidic device. Note that the SE image shown here has been recorded using a partial pressure of water vapor (~100 Pa) to help compensate against negative charging.

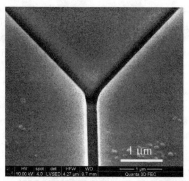

Figure 3. Nano-fluidic channels FIB-milled into quartz using automated electron beam charge control to suppress ion beam drift.

Ultra-high resolution S/TEM imaging & nanoanalysis

FIB specimen preparation techniques have been successfully used to generate ultra-thin lamellae for TEM and STEM, and there are now many examples in the literature. This approach can also be used to create ultra-fine, conical specimens suitable for atom probe experiments. One example is the determination of the architecture of quantum well structures, where time-of-flight atom probe microcopy can be used to generate a 3-dimensional reconstruction at the atomic level [5].

FIB milling is often performed using high energy Ga$^+$ ions (i.e., 30 keV), which can impart damage up to ~20 nm at the surface of a specimen such as silicon [6]. Given that both the upper and lower surfaces of a lamellar specimen carry this damage, it can easily be sufficient to impede quantitative high resolution S/TEM imaging and is especially problematic for analytical techniques such as electron energy loss spectroscopy (EELS), where the total specimen thickness should ideally be on the order of 30 nm or so. The advent of aberration-corrected S/TEM instruments capable of sub-angstrom resolution [7, 8] makes it ever more imperative that specimens are prepared to the highest possible standards and, clearly, methods are needed to minimize both ion implantation and surface damage if we are to obtain meaningful results. Amorphization damage in silicon and gallium nitride has been shown to be removed by chemical wet polishing after FIB milling [9, 10], but chemical polishing methods are material dependent and are difficult or impossible to use for complex multi-phase or multi-layered specimens. Broad beam ion milling is another approach to removing damage post-operatively [9-11]. However, perhaps a better solution would be to reduce the overall introduction of artefacts in the first instance, by FIB milling at lower ion beam energies. Recent advances in FIB column design allow us to go to beam energies as low as 500 eV whilst still maintaining a good beam profile. Hence, if a low energy ion beam is used for the final cleaning step in the specimen preparation process, any surface damage from high energy milling can be removed and replaced by a much smaller amount of damage. A study of silicon cleaned with a focused 2 keV Ga$^+$ ion beam shows sidewall amorphization to be as little as ~ 1 - 2 nm or so [6]. Whilst not eliminating the damage completely, the results are within acceptable limits for high-resolution imaging and analysis. The concept is illustrated schematically in figure 4, while figure 5 demonstrates that sub-angstrom resolution can be achieved with a FIB-prepared silicon specimen.

Little crystalline   >80% crystalline   >90% crystalline
material left   material left   material left

Figure 4. Schematic diagrams to illustrate the effect of FIB milling a thin, lamellar specimen at successively lower ion beam energies. Left-to-right: 30 keV, 5 keV and 2 keV, respectively. Damage and ion implantation reduces with beam energy, to yield specimens suitable for high-resolution imaging and analysis.

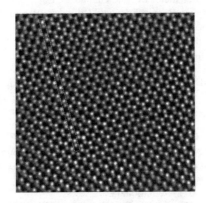

Figure 5. C_s-corrected TEM image of silicon, showing sub-angstrom resolution. The lamellar specimen was prepared using a focused Ga^+ ion beam. For the final cleaning step, an energy of 2keV was used, resulting in a very high-quality specimen.

CONCLUSIONS

The combination of FIB and SEM in one instrument enables us to perform a multitude of tasks. Rapid FIB prototyping and electron beam lithography allows us to choose the most suitable patterning techniques and strategies, offers a means for delivering prototypes for application testing long before batch processes are established and, in parallel, suggests scenarios for nano-fabrication in larger volumes. Surface cleaning using focused low energy gallium ions yields high-quality, site-specific specimens for ultra high-resolution studies such as S/TEM and atom probe microscopy, and techniques for charge control during milling, deposition and imaging are important steps in successfully applying FIB SEM methodologies to insulating materials.

ACKNOWLEDGMENTS

The help of the following colleagues is gratefully acknowledged: Francis Morrissey, David Wall, Steve Reyntjens, Bert Freitag & Tomas Vystavel.

REFERENCES

1.   Wilhelmi, O., Reyntjens, S., Roussel L., Stokes D.J. and Hubert, D.H.W. High Resolution Nanolithography using Focused Ion Beam Scanning Electron Microscopy (FIB SEM). (Manuscript in preparation).
2.   Morrissey, F., Reyntjens, S, Nakahara, K and Jiao, C., *Methods for Structuring and Prototyping on a Nanoscale Using a DualBeam.* Microsc. Microanal., 2005. 11(Suppl 2)
3.   Anzalone, P.A., Mansfield, J.F. and Giannuzzi, L.A., *DualBeam Milling and Deposition of Complex Structures Using Bitmap Files and Digital Patterning.* Microsc. Microanal., 2004. 10 (Suppl 2): p. 1154-1155.
4.   Stokes, D.J, Vystavel, T and Morrissey, F (2006) *Focused Ion Beam (FIB) Milling of Electrically Insulating Specimens Using Simultaneous Primary Electron and Ion Beam Irradiation.* Journal of Physics D: Applied Physics, 40, 874-877.
5.   Galtrey, M.J., Oliver, R.A., Kappers, M.J., Humphreys, C.J., Stokes, D.J., Clifton, P.H. and Cerezo, A (2007) Three Dimensional Atom Probe Studies of an $In_xGa_{1-x}N/GaN$ Multiple Quantum Well Structure: Assessment of Possible Indium Clustering. Applied Physics Letters 90 (6).
6.   Giannuzzi, L.A., Guerts, R. and Ringnalda, J, *2 kV Ga$^+$ FIB milling for reducing amourphous damage in silicon.* Microsc. Microanal., 2005. 11(Suppl 2): p. 828-829.
7.   Kujawa, S., Freitag, B. and Hubert, D.H.W., *An aberration Corrected (S)TEM Microscope for Nanoresearch.* Microscopy Today, July 2005: p. 16-21.
8.   Hubert, DHW (2007) International Lab News.
9.   N.I. Kato et al., J. Vac. Sci. Technol. A 17(4) (1999) 1201.
10.   J.R. Jinschek et al., *Microsc. Microanal.,* 10(supp 2) (2004) 1142.
11.   R.M. Langford and A.K. Petford-Long, *J.Vac.Sci.Technol.,* A 19(3) (2001) 982.

Mater. Res. Soc. Symp. Proc. Vol. 1020 © 2007 Materials Research Society   1020-GG01-07

Spatially Resolved Characterization of Plastic Deformation Induced by Focused-Ion Beam Processing in Structured InGaN/GaN Layers

R. Barabash[1,2], G. Ice[1], R. Kroger[3], H. Lohmeyer[3], K. Sebald[3], J. Gutowski[3], T. Bottcher[3], D. Hommel[3], W. Liu[4], and J.-S. Chung[1,5]

[1]Materials Science and Technology, Oak Ridge National Laboratory, One Bethel Valley Road, Oak Ridge, TN, 37831-6118

[2]University of Tennessee, Knoxville, TN, 37996

[3]Institute of Solid State Physics, Bremen, Germany

[4]Advanced Photon Source, Argonne, IL, 60439

[5]Soongsil University, Seoul, Korea, Republic of

ABSTRACT

Polychromatic X-ray microbeam analysis (PXM) results of structural changes caused by FIB machining in nitride heterostructures are discussed in connection with micro-photoluminescence (μ-PL), fluorescent analysis, scanning electron (SEM) and transmission electron microscopy (TEM) data. It is shown that FIB processing distorts the lattice in the InGaN/GaN layer not only in the immediate vicinity of the processed area but also in surrounding volumes. A narrow amorphized top layer is formed in the direct ion beam impact area.

INTRODUCTION

Focused-ion beam (FIB) machining is a promising technique for the realization of novel micro- and nanostructured optoelectronic devices- especially for the important nitride materials system [1], which is difficult to process by conventional etching techniques. For example, there is intense recent research directed at the application of FIB machining to create vertical semiconductor microcavities that possess three-dimensionally confined optical modes for surface-emitting lasers. Because these and other potential optoelectronic devices are sensitive to defect distributions, there is a strong need to assess the damage induced on GaN and related materials by FIB machining.

Although widely used for TEM-sample preparation and for processing of edge-emitters, FIB is increasingly applied to prepare mesa-structures for single-dot spectroscopy and for the preparation of micropillars out of planar vertical-cavity surface-emitting lasers (VCSEL) samples. In such materials induced damage can be important. Of course the damage induced by FIB machining has been previously studied in Si [2] and other materials [3]. For the nitrides however, there are only a few reports with respect to possible ion damage induced by FIB machining and similarly few reports on the induced damage impact on mesoscale structure and optical properties [4]. Below we present the results of polychromatic X-ray microbeam analysis of the damage in InGaN layers structured with FIB. The results are complemented by fluorescence, SEM, TEM and micro-photoluminescence (μ-PL) analyses.

EXPERIMENTAL DETAILS

A FEI Nova 200 NanoLab FIB system was used to prepare reference structure samples consisting of InGaN/GaN multi quantum wells grown by metal-organic vapor phase epitaxy on GaN on sapphire templates (Fig. 1). Some samples were capped with 100nm SiO_2 layer before FIB processing. Both capped and uncapped samples were studied. Each structure consists of several trenches typically 2 μm wide and 20 μm long with varying distances between the trenches. Trenches were etched down to the sapphire substrate using 30 keV Ga-ions with ion-beam currents varying from 300 pA to 7 nA.

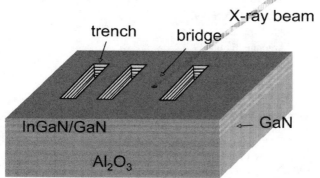

Figure 1. Schematic of the relative orientation of the X-ray microbeam (red spot) and the FIB processed trenches in the InGaN/GaN multilayer with trenches on sapphire.

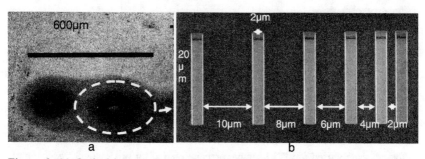

Figure 2. (a) Optical image of the structured region with several structures performed with different ion-beam currents in an uncapped sample; (b) Magnified SEM image of a typical FIB structure with varying distances between the trenches.

Several groups of trenches were processed for each sample. Two such groups in the middle of the shadowed area are shown in Fig. 2a. Typical FIB structures processed with FIB in the InGaN/GaN multi quantum wells on sapphire templates consisted of several bridges with varying distances between trenches (Fig. 2b). Further details of the FIB process can be found in Ref. [1].

Polychromatic X-ray microdiffraction (PXM) was used for spatially resolved structural investigations. PXM was performed with a focused ~0.5 μm diameter polychromatic synchrotron beam which hit the sample surface at 45° incidence (Fig. 1). The Laue diffraction patterns from both the InGaN layer and the sapphire substrate were recorded with a charge-coupled device (CCD) area detector placed at an angle of 90° relative to the incident beam [5 - 7]. To get depth resolved information about the strain distribution in the GaN layers we performed measurements with the so called DAXM technique [5].

With DAXM, a three-dimensional map of the local crystallographic orientation and elastic strain tensor distributions in a sample can be determined. The three dimensional orientation distributions can be used to directly determine the dislocation tensor field. Further details on the experimental setup, the data collection and quantitative analysis of diffraction profiles in PXM and DAXM can be found elsewhere [5 - 7].

The samples were also characterized by scanning electron microscopy (SEM) and by orientation imaging microscopy (OIM) using the field emission microscope of a FEI NOVA 200 focused ion beam system.

RESULTS AND DISCUSSION

Severe damage was observed in the uncapped sample. Micro Photoluminescence (μ-PL) measurements of the uncapped sample show that μ-PL intensity starts to decrease within ~200 μm of the structures area. On the structure itself no μ-PL- signal could be detected independent of beam current. We assume a chemical reaction between the Ga cloud and the uncapped GaN surface (Compare shadows around each structure at Fig.2a) may account for this loss of PL intensity.

In the case of the capped samples there was no change of the μ-PL intensity in the area surrounding the trenches. Therefore it was possible to measure the change of the PL-intensity at the different bridges (Fig. 3).

A scan of μ-PL intensity across the 10 μm- bridge (Fig. 3) starts in the middle of the trench and goes to the middle of the next trench. For each 10 μm - bridge the spectral position of the μ-PL intensity changes. It increases, stays practically constant in the middle of the bridge and decreases again when the laser spot reaches the end of the bridge. The reliable part of the measurement lies between two vertical lines, which correspond to the situation when the whole laser-spot is on the bridge. The change of the spectral position of the GaN signal may be caused by stress relaxation.

TEM analyses in the vicinity of the trenches shows a high density of dislocations and a narrow (~ 2nm) amophized top layer.

To investigate the interaction of stress relaxation and FIB induced structural damage, we used polychromatic x-ray microdiffraction (PXM) method. A typical PXM image for structured InGaN/GaN layers on sapphire substrate is shown in Fig. 4a. Due to the large penetration depth of the x-rays, the sapphire reflections are superimposed on those from the GaN.

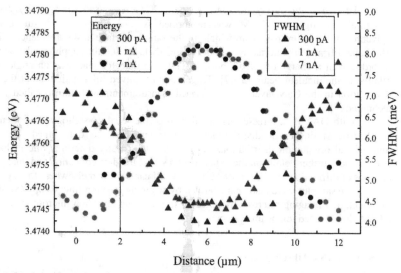

Figure 3. A scan of PL intensity across the 10 µm- bridge for the structures produces with the different beam currents for the capped sample.

PXM provides information about the local shear strain since a lattice distortion causes streaking of the Laue reflections. Lattice curvature can result from both FIB induced elastic as well as plastic distortion of the lattice. In this study, PXM images were obtained in geometry with the trench direction perpendicular to the beam. The total Laue pattern contains zone lines formed by InGaN and sapphire Laue reflections (Fig. 4a). The small dashed rectangle marks the area of interest containing the InGaN and the sapphire reflections.

Fluorescent analysis (Fig. 4b) was performed prior to the PXM analysis to find the exact location of the trenches which are shown as purple horizontal rectangles on the image. This allowed us to choose the correct locations for microbeam probing of the sample along vertical lines passing through the centers of the trenches (dashed line A1-A2) and through the edges of the trenches (dashed lines B1-B2 and C1-C2). The step size between neighboring probe locations along each line was 2µm. The starting probing location is indicated by an "x" on each line.

For PXM measurements the microbeam was scanned both parallel and perpendicular to the trenches to get a 2D map of the structural changes in the trenched region. A portion of the 2D map shown in Fig.5 corresponds to the region containing one upper trench marked by the dashed rectangle in Fig. 4b. Laue patterns were recorded at 50 different sample positions. Due to the small size of the microbeam (0.5 µm) it was possible to obtain spatially resolved data from different locations around the trenches. Because the high-energy (8-25 keV) X-ray beam penetrates through the InGaN/GaN film and probes the sapphire substrate as well, both the substrate and overlayer contribute to the observed Laue patterns. The sapphire reflections do not change position with sample translation and were used as a reference to determine changes in the local orientation of the

InGaN/GaN layer relative to the substrate (Fig. 5). The PXM results show that FIB etching distorts the lattice in the InGaN/GaN layer not only in the immediate trench region but also in the surrounding area. Lattice planes were curved with the curvature radius dependent on the distance from the trench, FIB current and the capping layer.

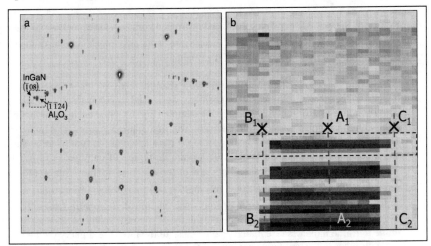

Figure 4. (a) Total Laue pattern with overlapped InGaN (round spots) and sapphire (elongated spots) reflections. The dashed rectangle marks the area of interest containing the InGaN and the sapphire reflections. (b) Fluorescent 2D map showing relative positions of trenches (purple rectangles) and X-ray microdiffraction probing locations (along the lines A1-A2, B1-B2, C1-C2).

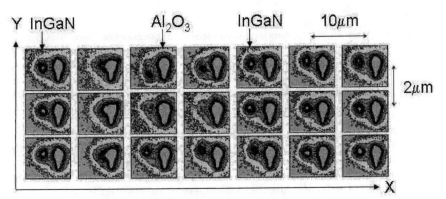

Figure 5. 2D map demonstrates lattice rotations in the InGaN layer relative to substrate.

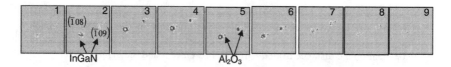

Figure 6. Several 3D depth-resolved (reconstructed) images corresponding to the region with $(\bar{1}08)$ and $(\bar{1}09)$ InGaN Laue spots and two week sapphire Laue spots demonstrate that there is a gradient of the orientation with depth.

With the DAXM technique, the wire shadows part of the diffracted intensity. Images collected with different wire positions were used to reconstruct the Laue images corresponding to different depth inside the sample (Fig. 6). Here each depth increment is 1µm. Depth is shown at each frame in µm. The first frame is closest to sample surface. The series of the reconstructed images in Fig.6 were taken at the first location shown by an "x" on the line A1-A2 in the Fig. 4b within the region of continuous film. Only the $(\bar{1}08)$ and $(\bar{1}09)$ InGaN Laue spots are observed close to the surface (1st and 2nd frames). With increasing depth the $(\bar{1}08)$ and $(\bar{1}09)$ InGaN Laue spots move indicating a change of orientation. At a depth of four - five microns the sapphire Laue spots appear (weak spots to the right of the InGaN spots at the 5th frame). Positions displacement from frame to frame indicates orientation change close to the interface. Finally the InGaN Laue spots disappear. A quantitative analysis of the orientation change with depth and with the distance from the trench is underway.

PXM data suggest the formation of dislocation walls. Although the defect structures clearly self-organize during FIB machining, further measurements will be required to determine whether the defects arise primarily from the inherent misfit dislocations of the original structure or from defects introduced during FIB machining.

CONCLUSIONS

- PXM and DAXM analysis have been used in a 3D spatially-resolved investigation of the local lattice orientation and defect microstructure in FIB structured InGaN layers grown on sapphire substrate.
- The PXM results show that FIB etching distorts the lattice in the InGaN/GaN layer not only in the immediate trench region but also in the surrounding area. Lattice planes become curved with curvature radii dependent on the distance from the trench, FIB current and the capping layer.
- The TEM analysis in the vicinity of the trenches shows a high density of dislocations and a narrow (~2 nm) amophized layer on top, which could be due to direct surface damage by the FIB beam or due to re-deposition.
- FIB induces structural damage of the GaN lattice is discussed in terms of the formation of disconnections causing long-range rotations in the layer.

ACKNOWLEDGEMENT

The research was supported in part by the U. S. Department of Energy, Division of Materials Science and Engineering through a contract with the Oak Ridge National Laboratory. Oak Ridge National Laboratory (ORNL) is operated by UT-Battelle, LLC, for the U.S. Department of Energy under contract DE-AC05-00OR22725. Data collection with PXM has been carried out on beamline ID-34-E at the Advanced Photon Source, Argonne IL. The APS is supported by the U.S. DOE, Basic Energy Sciences, Office of Science under contract No. W-31-109-ENG-38. This work was further supported by the Deutsche Forschungsgemeinschaft under Contracts No. HE 2827/5-1 and HO 1388/25-2.

REFERENCES

1. H. Lohmeyer, K. Sebald, J. Gutowski, R. Kröger, C. Kruse, D. Hommel, J. Wiersig, F. Jahnke, *Eur. Phys. J.* B **48**, 291-294 (2005).

2. Maaß, R., Grolimund, D. , Van Petegem, S., Willimann, M. Jensen, M., Van Swygenhoven, H., Lehnert, T. Gijs, M.A.M., Volkert, C.A., Lilleodden, E.T., & Schwaiger, R. *Appl. Phys. Lett.* **89**, 151905 (2006).

3. J. R. Greer, W.C. Oliver, W.D. Nix, *Acta Mater.*,**53**,1821-1830 (2005)

4. S.X. Jin, J.Z. Li, J.Y. Lin, H.X. Jiang, *Appl.Phys.Lett.* **76**, 631 (2000)

5. R.I. Barabash, G.E. Ice, (2005). *"Microdiffraction Analysis of Hierarchical Dislocation Organization"* In: Encyclopedia of Materials: Science and Technology Updates, Elsevier, Oxford. 1-18

6. R.I. Barabash, C. Roder, G.E. Ice, S. Einfeldt, J.D. Budai, O.M. Barabash, S. Figge, D. Hommel, *J Appl. Physics*, **100**, 053103-1 -053103-11 (2006)

7. R.I. Barabash, G.E. Ice, W. Liu, S. Einfeldt, D. Hommel, A.M. Roskovski, R.F. Davis, *Physica Status Solidi* (a), **202**, 5, 732-738, (2005)

Mater. Res. Soc. Symp. Proc. Vol. 1020 © 2007 Materials Research Society   1020-GG02-01

Dynamics of Ion Beam Stimulated Surface Mass Transport to Nanopores

David P. Hoogerheide[1], and Jene A. Golovchenko[2]
[1]Department of Physics, Harvard University, Cambridge, MA, 02138
[2]Department of Physics and School of Engineering and Applied Sciences, Harvard University, Cambridge, MA, 02138

ABSTRACT

We explore the ion beam-induced dynamics of the formation of large features at the edges of nanopores in freestanding silicon nitride membranes. The shape and size of these "nanovolcanoes", together with the rate at which the nanopores open or close, are shown to be strongly influenced by sample temperature. Volcano formation and pore closing slow and stop at low temperatures and saturate at high temperatures. Nanopore volcano size and closing rates are dependent on initial pore size. We discuss both surface diffusion and viscous flow models in the context of these observed phenomena.

INTRODUCTION

The advent of solid state nanopores as single-molecule detectors [1] has highlighted the importance of characterizing novel methods for fabricating nanostructures. Ion beam sculpting [2] is a robust method for making solid state nanopores, but the processes involved in the fabrication process remain poorly understood. While many potentially relevant processes, such as surface diffusion [2], sputtering [3–4], anisotropic deformation via thermal spikes [5–6], ion-enhanced viscous flow [6–8], radiation-induced stresses [9], and combinations of the above [10] have been studied with MeV beams or with crystalline targets, there is little data available regarding interactions of lower-energy keV beams with amorphous materials such as those found widely in silicon-based technology. In particular, it is unclear how matter transport occurs over large length scales when irradiated with low-energy beams that deposit their energy within only a few nanometers of the surface.

In this paper, we discuss several previously unreported features of nanopore fabrication using keV ion beams and discuss their ramifications with respect to various extant models for the observed matter transport.

EXPERIMENT

Samples were fabricated from low-stress amorphous silicon nitride grown by low-pressure chemical vapor deposition (LPCVD) on a silicon substrate. The LPCVD was performed at Cornell's Nanofabrication Facility and yields silicon nitride thin films with a nominal tensile stress of 180 MPa [11]. The stoichiometry is $Si_{3.5}N_4$, as determined by Rutherford backscattering. Freestanding square membranes 90 μm on a side and 500 nm thick were produced by lithographic patterning and a subsequent anisotropic etch of the underlying silicon substrate in KOH.

Starting holes of the desired diameter were drilled at the center of the freestanding membrane, from the substrate side, by rastering a 10-nm diameter, 50 keV gallium ion beam produced by a FEI/Micrion 9500 focused ion beam (FIB) instrument. The resulting holes were conical in shape, opening towards the substrate side of the membrane with an apex angle of about 10 degrees, as determined by atomic force microscopy (AFM).

After fabrication, samples were stored in a nitrogen-purged dry box until used. To ensure consistency of results, samples not in the dry box for more than a day were stored in methanol or ethanol and blown dry with dry nitrogen immediately before use. Samples were sometimes transferred directly from the FIB to the sculpting chamber, in which case no cleaning procedures were employed.

Ion beam sculpting experiments that form "nanovolcanoes" and modify the size of the nanopore were run in a home-built ion sculpting apparatus described elsewhere [12]. The base pressure of the sculpting chamber was about 7×10^{-10} torr with a cold can surrounding the sample in operation. The sculpting beam was a 3 keV ion beam of Ar^+ with a spot size of ~ 0.2 mm, which is significantly larger than the freestanding silicon nitride membrane upon which it impinges. The changing nanopore hole size was monitored instantaneously during the sculpting process with a channeltron single-ion detector located behind the sample (see Figure 1a for a highly schematic view).

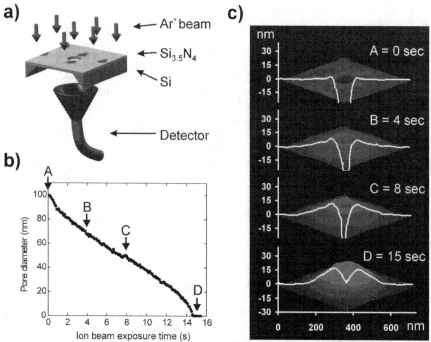

Figure 1. (a) Representation of ion sculpting experimental setup. (b) Pore shrinkage with time as deduced from the count rate. (c) Nanovolcano evolution.

To ensure consistency of results, each sample was baked at 80 °C for 5 minutes in the sculpting chamber prior to ion beam alignment. The beam was then aligned with the pore at a temperature of –30 °C, which is below the temperature at which most silicon nitride nanopores begin to close [2].

After ion beam sculpting the nanopores under the desired experimental conditions and recording the instantaneously changing rates at which ions were transmitted through the pore during the process, we studied the samples with an atomic force microscope. All topography analysis was performed on a Digital Instruments 3100 Nanoscope IIIa atomic force microscope (AFM) operating in tapping mode. Tips were Mikromasch NSC35A or NSC35B, with nominal tip radii of less than 10 nm and spring constants 7.5 N/m and 14 N/m, respectively. Images were processed with WSxM software [13].

EXPERIMENTAL RESULTS

A typical nanopore closing curve and volcano formation observed at different stages of the pore closing process are shown in Figure 1b-c. This pore was sculpted at 28 °C and at a flux of 4.4 $Ar^+/nm^2/sec$. The transmitted flux through the pore is monitored by the channeltron detector (Figure 1a), producing the diameter vs. time closing curve of Figure 1b. At various stages of closing, the sample was removed from the sputtering chamber and probed with an AFM, yielding the topographs in Figure 1c [14].

Temperature dependence

We have observed that both the instantaneous closing rate (*i.e.* the change in nanopore area with fluence or time, or the slope of an area vs. fluence curve) and volcano size vary significantly with sample temperature. Figure 2 depicts volcano shapes at varying temperature. All pores were closed from the same initial diameter (~ 100 nm) to the same final diameter (~ 8 nm).

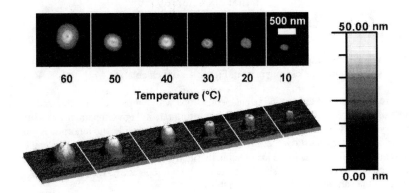

Figure 2. Nanovolcano sizes vary with temperature. Initial diameter is 100 nm for all pores, final diameter 8 nm.

It has previously been observed that pores that have been partially closed at high temperature will open at low temperature and continue closing at high temperature again [2]. Thus, a pore can be kept at approximately constant size by alternating temperatures at which the pore will open and close. Accordingly, to determine the temperature dependence of the instantaneous closing rate, the temperature of a single sample was cycled during closing. The pore was closed at 80 °C to about half its initial area. The temperature was then varied, alternating high (closing) and low (opening) temperatures to maintain a relatively constant area, as shown in Figure 3.

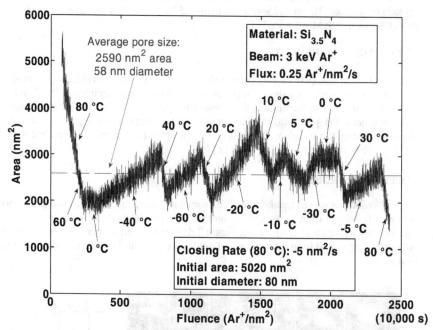

Figure 3. Nanopore ion sculpting signals vs. temperature. The red dashed line indicates the average pore size at which the closing rates in Figure 4 were calculated.

Figure 4 depicts the change in instantaneous slope with temperature, normalized to the closing rate of the pore at 80 °C. The saturation of closing rate at high temperature was first observed very recently by H. B. George and M. J. Aziz in silicon oxide and amorphous silicon nanopores [15], and our data show that the same effect occurs in silicon nitride. The temperature transition seems to indicate the presence of some thermally activated process, but it is not immediately clear why that process should saturate at high temperature. We present an explanation for the saturation behavior later in this paper.

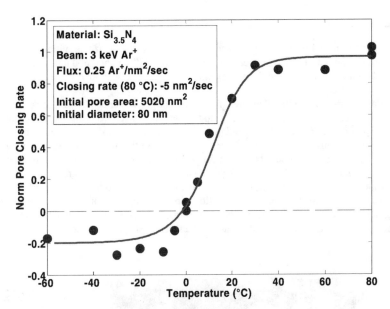

Figure 4. Nanopore ion sculpting rates vary with temperature. Rates are normalized to the area closing rate of the sample at 2590 nm^2 and 80 °C; negative rates indicate pore opening. The red curve is a fit to the diffusion model (discussed below).

The sensitivity of volcano size to changes in temperature in the transition region allows us to control the shape of the nanopore. Figure 5 shows three nanopores: one was closed at constant temperature, creating a standard nanovolcano; one at two different temperatures, producing a small volcano on top of a large one; and one at gradually decreasing temperatures, yielding a volcano with a smooth transition to the flat surface.

Figure 5. Ion sculpting of nanopores. Nanopores were closed at 0.9 Ar$^+$/nm^2/sec flux, 100 nm initial diameter, and temperature (a) constant 40 °C; (b) 60 °C to half initial area, then 20 °C; (c) temperature decreasing from 60 °C to 40 °C to 20 °C. Note the "volcano on a volcano" feature in (b). Heights are (a) 32 nm; (b) 28 nm; (c) 29 nm above the sputtered surface. All figures have lateral dimensions 1425 nm ° 1425 nm.

Initial size dependence

We also investigated the dependence of volcano shape and size on initial pore area. Four nanopores, of nominal initial diameters 100 nm, 200 nm, 400 nm, and 600 nm, were closed completely under identical conditions: temperature 80 °C, flux 0.25 Ar$^+$/nm^2/sec. The membrane thickness for these pores was 250 nm.

Pores with larger initial diameters form much larger volcanoes (see Figure 6) and thus require more material to shrink the pore by adding another "layer" to the volcano. Assuming mass transport occurs at a constant speed, pores with larger volcanoes should close more slowly at a given area, in this case 4000 nm^2. Therefore, we expect that for a given instantaneous pore diameter, the closing rate should be slower for pores that started larger. Indeed, we observe that pores with initial diameters larger than 200 nm close more slowly with increasing pore size (see Figure 7 and Figure 8). We do not have an explanation for the drop in closing speed for small initial diameters. In order to solidify the relationship between instantaneous volcano shape and closing rate, future work might include investigating closing at some final diameter and comparing the closing rates at that diameter to the volcano shape at that point.

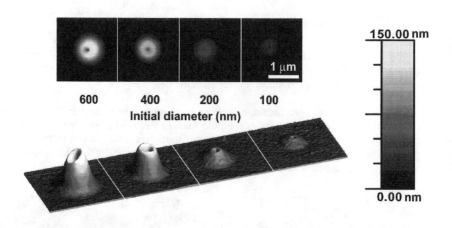

Figure 6. Initial area dependence of volcano size. All pores were closed at 80 °C and a flux of 0.25 Ar$^+$/nm^2/sec. Initial diameters are nominal.

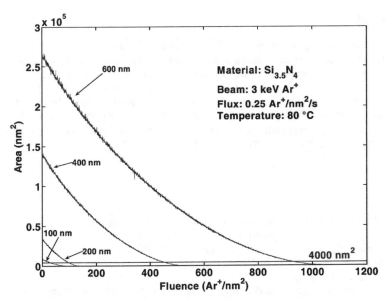

Figure 7. Dependence of ion sculpting behavior on initial nanopore area. Instantaneous slopes are compared at 4000 nm^2 in Figure 8.

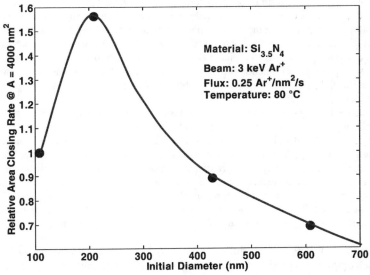

Figure 8. The closing rate at a given area relates to the initial size of the nanopore. The red curve is a guide to the eye. Initial diameters were determined by transmission electron microscopy (TEM).

DISCUSSION

In order to explain pore closing data, various models have been explored. Originally, Li *et. al.* [2] advanced a surface diffusion model in which beam-generated adatoms diffused on the surface until they were captured by surface traps, were immobilized on the pore rim, or were annihilated by the beam. The closing behavior in this model is characterized by the parameter X_m, the average distance an adatom travels before it is trapped or destroyed by the beam, or, equivalently, the size of a depletion region of adatoms around the pore. X_m is given by Equation 1, where l_{trap} is the average distance between surface traps, D is the surface diffusivity, σ is the cross-section for annihilation by the beam, and f is the beam flux.

$$\frac{1}{X_m^2} = \frac{1}{l_{trap}^2} + \frac{\sigma}{D} f \tag{1}$$

The surface diffusion model required modification in order to explain both the accretion of relatively vast amounts of material in the nanovolcanoes and the large distances from which this material appeared to be coming. Mitsui *et. al.* [14], after studying sculpting dynamics of arrays of nanopores, proposed an enhancement to the diffusion process resulting from large electric fields on the surface.

The surface diffusion model also lends itself readily to incorporation of temperature effects. Except for very large pores, the closing rate is proportional to X_m. It is then clear from Equation 1 that an Arrhenius form of the surface diffusivity will yield the sigmoidal behavior observed for the temperature dependence in Figure 4. At high temperatures, where diffusion is significant, the average "lifetime" of an adatom is limited by the trap concentration l_{trap}. At low temperatures, the adatom is diffusion-limited and is more likely to be annihilated by the beam than to fall into a trap or reach the pore periphery. If at reduced temperatures diffusion is thermally suppressed sufficiently, the removal of material from the pore periphery by sputtering dominates, and the pore opens. A typical value for the average trap separation distance is $l_{trap} \sim$ 65 nm. The data are not sufficiently precise to extract reliable Arrhenius parameters; we only report that the surface diffusivity varies from about 1 to 10^3 nm^2/s over the temperature range of the experiment.

The primary weakness of the surface diffusion model lies in explaining the size and shape of the volcanoes; even the possible enhancement by charging seems inadequate to explain this extraordinary effect in detail. These concerns with the diffusion model have led some to speculate about the contribution of ion-enhanced viscous flow such as that seen for MeV beams [6–8]. In this case the viscosity of the irradiated surface is reduced considerably by the deposition of energy from the beam, causing flow that responds to intrinsic or ion-induced stresses in the surface. It is natural to explain volcanoes as buckling due to stress relief or the result of a collective flow toward the pore. The dependence on initial pore size in this scenario may arise from differences in stress from the initial FIB milling or from the curvature of the pore.

The viscous flow model is difficult to quantify, however, and it is unclear how the observed temperature dependence would arise in such a model, where the important terms are not strongly temperature-dependent at room temperature [10]. The "thermal spike" usually associated with ion impingement at high energy should not be sensitive to the relatively small changes in surface temperature in our experiments.

CONCLUSIONS

Despite the new data presented in this paper, it is still inconclusive whether surface diffusion, viscous flow, a combination of these, or some other mechanism is responsible for the large matter transport observed in the ion sculpting process. We have introduced constraints on possible models based on temperature and initial size dependence and have demonstrated how these effects can be used to control the shape of a nanovolcano. Understanding the material transport mechanism and its interplay with electric fields and the underlying material properties could greatly aid in the precise fabrication of novel nanostructures.

Ion sculpting of nanopores also presents a powerful platform for studying the effects of keV ion beams on amorphous surfaces. Despite increasing interest in nanofabrication using lab-scale ion beams, a clear microscopic understanding of the interaction of keV ion beams with amorphous surfaces remains elusive. It is difficult to imagine modeling these interactions without elucidating and accounting for the mechanisms that create nanovolcanoes. Any answer to the question of matter transport under ion irradiation, be it individual adatom diffusion, collective flow, or a combination of the two, further influenced by implanted charge, must be able to explain the dramatic presence and morphology of nanovolcanoes.

ACKNOWLEDGMENTS

We wish to acknowledge technical and design assistance from Marc Gershow and valuable discussions with Dr. Bola George and Professor Mike Aziz. D. Hoogerheide is supported by an NDSEG graduate fellowship.

REFERENCES

1. J. Li, M. Gershow, D. Stein, E. Brandin, and J. A. Golovchenko, Nat. Mater. **2**, 611 (2003).
2. J. Li, D. Stein, C. McMullan, D. Branton, M. J. Aziz, and J. A. Golovchenko, Nature (London) **412**, 166 (2001).
3. H. Gnaser, *Ion Irradiation of Solid Surfaces* (Springer, Berlin, 1999).
4. M. A. Makeev and A.-L. Barabási, Appl. Phys. Lett. **71**, 2800 (1997).
5. M. L. Brongersma, E. Snoeks, T. van Dillen, and A. Polman, J. Appl. Phys. **88**, 59 (2000).
6. E. Snoeks, T. Weber, A. Cacciato, and A. Polman, J. Appl. Phys. **78**, 4723 (1995).
7. H. Trinkhaus and A. I. Ryazanov, Phys. Rev. Lett. **74**, 5072 (1995).
8. C. C. Umbach, R. L. Headrick, and K.-C. Chang, Phys. Rev. Lett. **87**, 246104 (2001).
9. Y.-R. Kim, P. Chen, M. J. Aziz, D. Branton, and J. J. Vlassak, J. Appl. Phys. **100**, 104332 (2006).
10. K. Otani, X. Chen, J. W. Hutchinson, J. F. Chervinsky, and M. J. Aziz, J. Appl. Phys. **100**, 023535 (2006).
11. "CNF - Chemical Vapor Deposition Capabilities" (http://www.cnf.cornell.edu/cnf_process_tf_cvd.html).
12. D. M. Stein, C. J. McMullan, J. Li, and J. A. Golovchenko, Rev. Sci. Instrum. **75**, 900 (2004).

13. I. Horcas, R. Fernandez, J. M. Gomez-Rodriguez, J. Colchero, J. Gomez-Herrero, and A.M. Baro, Rev. Sci. Instrum. 78, 013705 (2007).
14. T. Mitsui, D. Stein, Y.-R. Kim, D. Hoogerheide, and J. A. Golovchenko, Phys. Rev. Lett. **96**, 036102 (2006).
15. H. B. George, Ph. D. thesis, Harvard University, 2007.

Mater. Res. Soc. Symp. Proc. Vol. 1020 © 2007 Materials Research Society   1020-GG02-03

Ion Beam Lithography for Nano-scale Pattern Features

John E.E. Baglin, Andrew J. Kellock, and Jane E. Frommer
IBM Almaden Research Center, 650 Harry Road, San Jose, CA, 95120

ABSTRACT

With the expected availability of new tools for creating patterned ion beams containing few-nanometer sized features, it is important to examine the fidelity of registering such patterns in a receiving medium, such as the photoresist layer in a lithographic fabrication process. In this paper, we report experiments exploring the characteristics of ion beam patterning of polymethylmethacrylate (PMMA) and polystyrene (PS) coatings on silicon substrates, with respect to their response as positive / negative resists to patterned low-energy H^+, He^+ and Ne^+ beams. We examine by atomic force microscopy (AFM) the feature profiles thus created after solvent development of the polymer layers, and we examine the dependence of the polymer response upon ion species and fluence. Reasonable feature profiles are readily obtained in fluence ranges around 10^{13} ions/cm^2. Proximity effects are shown to be negligible except after over-exposure at very high ion fluences. Granularity within the final pattern features is shown to be a potential concern for high energy, light ion irradiations. Optimization of feature geometries is clearly possible by appropriate selection of ion species, energy and fluence to suit the receiving medium.

INTRODUCTION

In view of their freedom from interference patterns, their minimal proximity effects in complex pattern lithography, and their adaptability for parallel processing, ion beams are expected to reach beyond the current manufacturing technologies based on e-beams or UV lithography. Achieving feature resolution of only a few nanometers, and offering highly flexible and reliable process throughput with long range pattern coherence, ion beams will enable fulfillment of Moore's Law for semiconductor systems, and similar advances for high density magnetic storage media. Prototype projection ion beam tools, designed for parallel transfer of a complete large-area pattern with demagnification factors between 4 and 20, have demonstrated the robustness, precision, and throughput needed to make this approach economically attractive [1-4]. Programmable masks and higher demagnification factors are expected to further enhance the scope of applications in the future.

It is therefore timely to explore the optimization, and physical limits, of spatial resolution in registering an ion beam pattern in materials suitable for the role of lithographic resists. We report an initial experimental study of the sharpness of pattern registration in thin films of polymethylmethacrylate (PMMA) or polystyrene (PS) exposed to low energy ions patterned by means of a contact stencil mask.

EXPERIMENT

Polymer resist coatings

In these experiments, we compared the performance characteristics for PMMA and PS. The linear PMMA polymer structure is readily broken by chain scission events caused by the passage of energetic ions, leading to enhanced susceptibility to a solvent developer, while response of the PS structure is more complex [5]. Both polymers are also expected to display crosslinking as a result of bonds broken during ion irradiation. Thus PMMA initially functions as a positive resist with a suitable solvent developer, while the onset of crosslinking at higher doses, generally accompanied by escape of volatile products, converts the polymer into an insoluble material, which then serves as a negative resist. PS, however, appears to function exclusively as a negative resist under ion irradiation. [5,6].

Samples of PMMA were prepared by spin-coating from a solution in isopropyl alcohol onto silicon substrates, followed by a post-bake in air for 5 minutes at 120°C to remove solvent residues. Coatings of thickness ~60 nm were obtained. PS samples were spin-coated from toluene solution, at ~80 nm thickness, with a 5-minute 100°C bakeout.

Procedures

For each ion beam exposure, a contact stencil mask was used. The mask consisted of silicon, 10 μm thick, containing a multiple-pitch array of 1 μm diameter holes (Fig. 1), and supported as a ~1 mm^2 window in its parent Si wafer. The masks were fabricated from a standard SOI (silicon-on-insulator) wafer, using standard silicon photolithography and etching [7]. Under SEM examination the transmission holes exhibited smooth, sharply defined walls. Samples were irradiated at room temperature, in vacuum $\leq 10^{-7}$ Torr. The 3 mm beam spot (with divergence $\leq 3 \times 10^{-4}$ msr) was asynchronously rastered to assure uniform exposure over the mask. Ion fluences were calibrated using a direct Faraday cup. Exposures were made for a series of fluences, ranging from 10^{12} ions/cm^2 to 5 x 10^{16} ions/cm^2. For consistency of coating and development, each series of exposures was made in separate locations on a single wafer

 sample. Following such a series of exposures, each PMMA-coated wafer was immersed in IPA:MIBK(isopropyl alcohol : methyl isobutyl ketone) = 3:1 developer for >30 seconds. PS samples were immersed in toluene for development. The resulting topography in the exposed patterns was then imaged by AFM, and line scans across representative features were compared.

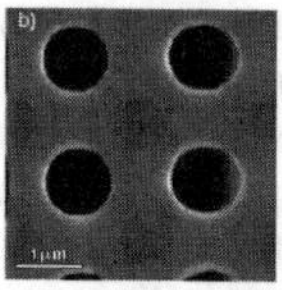

Figure 1. SEM image of a small section of the patterned stencil mask. (Hole diameters ~1 μm).

Ion species and energies

The following ion beam species and energies were selected in order to explore the effects of some significant differences in their dominant interaction modes with the polymers: H$^+$ (300 keV), He$^+$ (10 keV) and Ne$^+$ (20 keV). In all three cases, the ion's projected range extends well beyond the resist layer, and the <u>total</u> rates of energy deposition (eV per Å) in the PMMA and PS are reasonably constant through the polymer layer from the surface to the substrate. However,

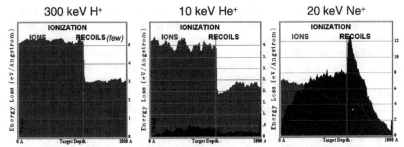

Energy deposited by ionization (eV/Angstrom) from primary ions and recoils in 60nm PMMA on silicon

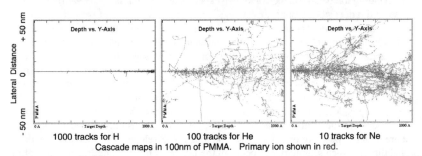

1000 tracks for H   100 tracks for He   10 tracks for Ne
Cascade maps in 100nm of PMMA.   Primary ion shown in red.

Figure 2. Energy deposition in 60 nm PMMA on silicon, and cascade maps for PMMA, [8].

as shown in Fig. 2, simulations made using the Monte Carlo code SRIM [8] display the following distinctly different features that might affect the response of the resist. The H^+ energy deposition in the polymer film occurs entirely by means of ionization. Also, in the near-absence of scattering of the fast primary ion, the direct interaction cascade is very narrow. The He^+ energy deposition in the layer is also nearly all by direct ionization processes; however, the He^+ cascade volume is relatively broad. The Ne^+ cascade is also very broad; however, for Ne^+, a substantial fraction of the energy deposited by ionization is derived from host atoms that are recoiling after ballistic scattering events. The significant number of displacement collisions produced by the Ne^+ ions suggests the possibility of a different damage mechanism that could modify the response of the polymer, and perhaps promote other reactions such as crosslinking. Rates of sputtering loss are negligible for all of these ions and energies.

Evaluation of pattern feature profiles

Ideally, an ion beam defined by a cylindrical hole in a transmission mask such as that of Figure 1 would modify a perfect right cylinder of resist extending from the surface to the substrate. After development, a positive resist would produce a cylindrical well in that space, while a negative resist would generate a free standing cylinder. In practice, the sharpness and steepness of the side walls of such generated features will define the minimum size and pitch of features achievable in a realistic patterning process. That sharpness and steepness of the feature side walls in resist may depend upon a variety of factors such as the choice of ion beam parameters, minimized ion scattering in the mask, and the choice of resist itself.

It is a primary purpose of this experiment to explore the quality of the wells or cylinders generated in a commonly used thickness of simple polymer resists, seeking especially the parameters that result in the steepest feature side walls. For this purpose, we have adopted as one metric for pattern fidelity the difference between the critical dimension of feature width at the resist surface (CD_{top}) and that at the substrate (CD_{bottom}). In order to obtain CD_{top} from an observed cross section, each side wall is characterized by a straight line of representative slope, and CD_{top} is defined as the separation of those lines where they meet the resist surface plane. Similarly, CD_{bottom} is defined as the separation of those side wall lines where they meet the substrate plane. As a quantity that serves to describe the extent of spreading of one side wall, we identify "edge resolution" as $|(CD_{top} - CD_{bottom})| / 2$. This "edge resolution" is thus a rough indicator of the minimum feature size that could be patterned successfully using the corresponding ion beam, fluence, and resist species and thickness. This metric is rudimentary, of course, and it does not account for details such as rounded corners or pattern granularity.

The fluence dependence of the profiles is of substantial interest, since it not only gives some indication about the critical intrinsic mechanism of ion-resist interaction, but it also indicates the breadth of processing window for practical usage. In the case of PMMA, it is especially interesting to observe the crossover between positive-resist and negative-resist performance, for fluences typically around 10^{14} ions/cm^2. Another important criterion for high quality in PMMA positive lithography is the completeness of removal of exposed material. An undissolved residue, either granular or continuous, would be highly undesirable. However, due to the stochastic nature of the ion interactions, we should expect to find local patches of resist that are under-exposed or over-exposed, and thus not developed, especially at low fluences. The probability for such granularity to occur in the feature is much higher for the case of the narrow H^+ tracks than it is for the more dispersed cascades of heavier ions. However, the very high fluences required for complete coverage of the feature area by the narrow tracks may then increase the risk of over-exposed spots that are not removed by developer. As we shall see, the issue of such granularity can be very significant.

Proximity effects

A limiting concern for standard lithography involving full-pattern transfer is the spurious effect of laterally-scattered energetic electrons or photons that contribute to the exposure budget of neighboring resist features. Programmed compensation for such "crosstalk" is possible, but complicated, and increasingly problematic for high density patterns at a few-nm scale. Due to their efficient energy transfer in the resist, low energy ion beams should be much less likely to cause proximity effects than energetic electron beams or photons, that require large input fluences, and whose lateral scattering range can be large. In order to identify fluence ranges where proximity effects may begin to appear, we extended the range of test fluences to very high values. As we shall see, such fluences were well outside of the dose range required for clean feature definition.

RESULTS

Figures 3 (a) and (b) show typical AFM images, together with typical placement of cross-sectional lines, for patterns obtained from PMMA after exposure to 10 keV He$^+$, and subsequent development.

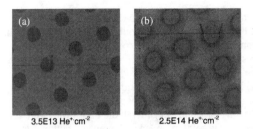

3.5E13 He$^+$cm^{-2} 2.5E14 He$^+$cm^{-2}

Figure 3. Typical AFM images for patterns obtained in PMMA corresponding to 10 keV He$^+$ exposure at (a) 3.5 x 10^{13} ions/cm^2 (b) 2.5 x 10^{14} ions/cm^2. [Each image 10 μm x 10 μm; z-scale 150nm]

Figures 4 (a), (b) and (c) show the profiles of the best-defined cylindrical wells obtained for PMMA, corresponding to selected fluences of Ne$^+$, He$^+$ and H$^+$. The "edge resolutions" for

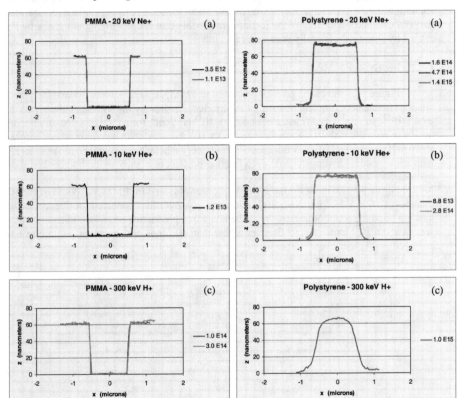

Figure 4. Typical feature profiles obtained as AFM cross sections, for PMMA at 'best' fluences of Ne$^+$, He$^+$, H$^+$.

Figure 5. Typical feature profiles obtained as AFM cross sections for PS at 'best' fluences of Ne$^+$, He$^+$, (and an insufficient, but high, fluence of H$^+$).

these features, averaged from several such measurements, were 25 nm for Ne⁺, 36 nm for He⁺,
and 41 nm for H⁺ (with typical reproducibility of ± 5 nm). The bottom of the exposed well made
with Ne^+ is apparently clean, and free from asperities. The He^+ exposure displays minor
evidence of residual asperities, and the residual granularity from H^+ exposure is also evident,
both at low fluences and at large fluences.

Figures 5 (a), (b) and (c) show typical profiles of free standing cylinders obtained for PS
after choosing the "best" fluences for Ne^+, He^+ and H^+. The corresponding "edge resolution"
assessments were 36 nm (for Ne^+) and 30 nm (for He^+). In general, the PS profiles were less
well defined than those from PMMA, their edges being rounded at the top and bottom. In the
case of He^+ irradiation, granularity was evident, and prevailed even at higher fluences.The case
of H^+ irradiation of PS is peculiar. After exposure at 10^{15} H^+/cm^2, conversion of the PS was still
evidently incomplete, while the walls of the "cylinder" were already broadly spread, and "edge
resolution" could not be assessed. Even higher fluences did not display full conversion, although
they did produce a widening of the feature, presumably due to crosslinking of the nearby PS
caused by secondary, laterally scattered, radiation.

High Fluence in PMMA

Figure 6(a) shows the profiles generated in PMMA by exposure to 20 keV Ne^+, for a
series of fluences. Complete removal of the irradiated region by the solvent is already enabled
by 3×10^{12} Ne^+/cm^2. At a fluence of about 3×10^{13} Ne^+/cm^2, formation of a non-soluble pillar
begins, eventually stabilizing at a fluence of 3×10^{14} Ne^+/cm^2 with a column whose height is
about half the original PMMA thickness – presumably a heavily cross-linked amorphous carbon
with a low residual hydrogen content – a familiar product of high-dose ion bombardment of
hydrocarbon materials.

Figure 6(b) shows a similar series of profiles obtained for PMMA exposed to various
fluences of 10 keV He^+ ions. For He^+ irradiation, sharp profiles appeared for fluences of 10^{13}
ions/cm², and the onset of negative resist behavior and production of an insoluble pillar were
seen at about 10^{14} ions/cm². At fluences higher than 10^{14} ions/cm², however, a complex lateral

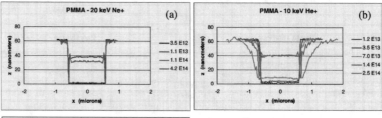

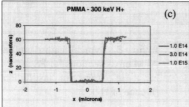

Figure 6. Typical feature profiles
recorded by AFM on PMMA after
exposure to various fluences,
including high fluences, of (a) Ne⁺,
(b) He⁺, (c) H⁺.

anomaly appeared at the boundary of the stencil image, in which the removal of some PMMA extended at the surface beyond the stencil feature by as much as several hundred nm, and a deep, narrow trench surrounded the pillar.

High fluence H^+ irradiation (Fig. 6(c)) displayed good "edge resolution", but concurrently the center of the feature showed the onset of negative tone performance.

The origin of the long-range effects from the He^+ irradiation is not clear at this point. The narrow trench surrounding the surviving well-defined core feature can be understood as a region just beyond the shadow of the mask edge, that may receive a small amount of damage (sufficient to respond as a positive resist), due to lateral straggling of incident ions and scattered particles in the ion cascades, and even penetration of a few scattered ions through the mask corner before entering the PMMA. However, the origin of the very broad halo near the surface, at high doses, is not clear. Scattered ions at these energies would not travel that far, and neither would secondary electrons created by these ions. In order to test the proposition that the effect is due to surface heating, we repeated these tests with beam current being reduced by an order of magnitude – but found no change in the resulting patterns. The physical source of these long range, high-dose anomalies remains to be determined. We do not believe that they are artifacts of our experimental procedure.

Fortunately, from a practical perspective, the ion fluence required to produce these long range proximity effects is at least one order of magnitude greater than that required for normal positive (or negative) resist exposure in PMMA.

DISCUSSION

These preliminary experiments serve to indicate that a feature "edge resolution" of 25 nm is not difficult to achieve, with quite broad process windows, using low-energy patterned beams of ions such as 20 keV Ne^+ with a resist coating of PMMA of thickness 60 nm. The typical fluence required for the positive resist regime is $\sim 10^{13}$ /cm^2, which in many ion-beam instruments would be delivered in a few seconds. This experiment reveals no unexpected effects at such fluences. Further improvement in registration may well be achievable by the use of ion projection systems or focused ion beam facilities, in which the contact stencil mask is replaced by a macroscopic master transmission mask or pattern generator, whose reduced, focused image selectively delivers its pattern of ions directly on the work piece.

The response of the PMMA to Ne^+, He^+ and H^+ ions as the ion fluence is progressively increased seems to be qualitatively consistent with a simple model in which individual, sparsely distributed ionization cascades produce local chain scission, and the entire exposed polymer becomes susceptible to dissolution in a solvent developer when the density of accumulated damage locations allows more or less continuous interconnected breakup paths. The feature-edge sharpness for this process will be limited by the lateral straggling of cascade particles within the polymer beyond the mask edge (approx. 25 nm for 20 keV Ne^+), and by the abruptness of the solubility threshold. This appears to be consistent with the observed 25 nm "edge resolution" for 20 keV Ne^+ in PMMA of thickness 60 nm. For higher fluences, the evolution of hydrogen and accompanying volatile hydrocarbons, and the consequent extensive crosslinking among residual structures [5,6] generates an insoluble core, whose sharp boundaries are surrounded by a narrow lateral-straggling ring, whose weak local exposure is still in the positive resist (soluble) regime. No potential proximity effects beyond the expected lateral straggling for a single cascade were actually observed for the Ne$^+$ irradiations.

In contrast, the lateral effects found for high-fluence 10 keV He$^+$ in PMMA are not easy to account for. Even at low fluence, the best "edge-resolution" observed was substantially wider than the predicted lateral straggling of the cascade. We speculate that these effects may in some way be a consequence of the low average local density of broken bonds over the broad straggling volume for He$^+$ ions, that will affect the kinetic competition between scissioning, rebonding and crosslinking processes prior to development. It is perhaps worth noting that similar lateral spreading was found for high fluence He$^+$ irradiation of the PS samples, again contrasting with the performance of Ne$^+$ irradiation. Since the effect prevails for both PMMA and PS, it seems that the explanation lies somewhere in the basic differences of cascade dispersion and mode of energy deposition between He$^+$ and Ne$^+$, rather than in the specific chemistry of the resist.

For H$^+$ irradiations, PMMA responded with good definition at low fluences, although here the "edge resolution" (~40 nm) was much larger than the lateral dimension of a single H$^+$ ionization cascade in the polymer, as modeled with SRIM. This implies that the observed side wall spreading is not attributable exclusively to lateral cascade straggling.

SUMMARY

In summary, these simple experiments serve to illustrate that ion beam patterning of polymer resists can readily achieve 25 nm "edge resolution" without proximity effects. Further, patterned ion beams offer the choice of many variables to suit the application; optimization of "edge resolution" and concurrent avoidance of granularity are important requirements for each case. In order to optimize these performance qualities systematically, a clearer model of the ion interaction processes in the polymer is needed, to enable realistic modeling of image formation. Such studies are now in progress.

Some granularity for the process remains an apparently unavoidable consequence of using high-energy light ions having very small cascade diameters. It remains to be studied in detail. Conceivably, the success of 20 keV Ne$^+$ in the present experiment (with its broad cascade) for PMMA is related to the multi-directional secondary recoil tracks deep in the resist layer, whose networks of damage can weaken the bulk material homogeneously in a way that could not be achieved by unidirectional, narrow tracks from energetic H$^+$ ions.

REFERENCES

1. Q. Ji, X. Jiang, L. Ji, Y. Chen, B. van der Akker and K.-N. Leung, "Novel Ion Beam Tools for Nanofabrication", Chapter 11 in Nanotech 2005, Volume 2, 703, NSTI (2005)
2. H. Loeschner, E.J. Fantner, R. Korntner, E. Platzgummer, G. Stengl, M. Zeininger, J.E.E. Baglin, R.. Berger, W.H. Brünger, A. Dietzel, M.-I. Baraton and L. Merhan, Mater. Res. Soc. Proc. 739, H1.3.1 (2003).
3. A. Dietzel, R. Berger, H. Loeschner, E. Platzgummer, G, Stengl, W.H. Bruenger and F. Letzkus, Advanced Materials 15, 1152 (2003).
4. J. Melngailis, A.A. Mondelli, I.L. Berry, R. Mohondro, J. Vac. Sci. Technol. B16, 97 (1998)
5. Eal H. Lee, "Ion Beam Modification of Polyimides", Chapter 17 in "Polyimides: Fundamentals and Applications", eds. M.H. Ghosh and K.Mittal, Marcel Dekker, New York (1996)

6. J.E.E. Baglin, A.J. Kellock, M.A. Crockett and A.H. Shih, *Nucl. Instrum. Meth.* **B64**, 469 (1992)
7. B.D. Terris, D. Weller, L. Folks, J.E.E. Baglin, A.J. Kellock, H. Rothuizen and P. Vettiger, *J. Appl. Phys.* **87**, 7004 (2000)
8. J.F. Ziegler, *"The Stopping and Range of Ions in Matter"*, SRIM-2006, http://www.srim.org

Mater. Res. Soc. Symp. Proc. Vol. 1020 © 2007 Materials Research Society   1020-GG02-04

Ion-Beam Assist Nano-Texturing of Films

Vladimir Matias, Chris Sheehan, and Alp T. Findikoglu

MPA-STC, Los Alamos National Laboratory, Mail Stop T004, Los Alamos, NM, 87545

ABSTRACT

We present an ion-beam based fabrication method for growth of single-crystal-like films that does not utilize epitaxy on single crystal substrates. We use ion-beam assisted texturing to obtain biaxial crystalline alignment in a film. This ion-beam assisted deposition (IBAD) texturing can be done on arbitrary, but smooth, substrates, including flexible polycrystalline metal tapes. With IBAD texturing of MgO and subsequent homoepitaxial growth we have demonstrated an in-plane mosaic spread FWHM as low as 2° and out-of-plane alignment of 1°. The deposition system we use includes reel-to-reel tape transport for a linear transport of substrate materials through the deposition zones. This allows for high-throughput experimentation via a linear combinatorial experimental design.

INTRODUCTION

Ion beam assisted deposition, or IBAD, where an ion beam is impinging on the sample during deposition of a film, is well known to be able to modify properties of films significantly. For a review of IBAD modifications see Ref. [1]. Among the many structural modifications that have been observed is the change in the film crystalline orientation. Figure 1 shows the relevant regime in the ion beam energy/current phase space. The regime is quite broad and encompasses a number of processes and materials that have been explored in the literature.

Of particular interest is the biaxial alignment of the crystallites in the film when an off-axis ion beam impinges during growth. Figure 2 shows a schematic of this process. The ion beam breaks the symmetry in the plane of the film and induces an in-plane crystalline alignment. This process is optimized for certain ion beam angles that correlate with an ion channeling direction or an otherwise preferred ion bombardment direction that minimizes radiation damage. The amount of crystalline texture attained depends on the details of the material and the process conditions. A key question in this field of research has been to understand how well the film can be oriented. Early work in this area was done at IBM, and it focused on IBAD of metal species, such as Nb [2]. In the early 1990's Iijima et al reported successful biaxial alignment of Y_2O_3-stabilized ZrO_2 [3]. This was followed in 1995 by the Stanford University group who demonstrated IBAD texturing of very thin MgO layers [4]. This last material and process is the topic of our research. The texturing for MgO happens most readily at the stages of nucleation and early film growth, where the film thickness is less than 10 nm. The ion beam for this nano-scale texturing is optimized at 45° to the substrate, so that it impinges in the <110> direction of the MgO crystal structure as the film grows in the (001) orientation.

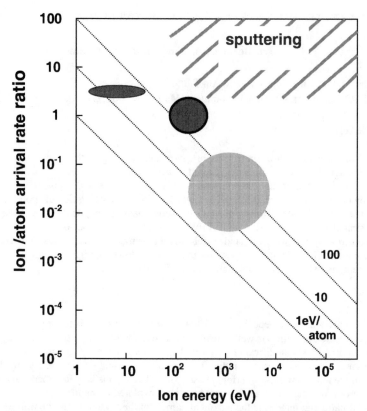

Fig. 1. Ion energy and normalized current plot for the regimes of IBAD. The red oval shows a regime used for improved step coverage of films. The large green circle is a regime for change of film stress. The blue circle with the outline is the regime reported for IBAD texturing. See Ref. [1].

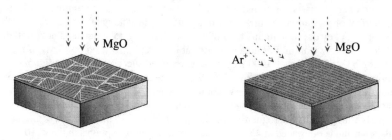

Fig. 2. Schematics showing the influence of an off-axis ion beam to obtain biaxial crystalline alignment. Left figure shows film deposition on a polycrystalline substrate where the film grains are randomly oriented. On the right the film deposition occurs concurrently with an ion-beam assist which orients the grains in the plane of the film, as well as out of the plane.

EXPERIMENT

A schematic of our deposition system is shown in Fig. 3. Reel-to-reel tape transport is used for a continuous feed of substrate. We use an electron-beam evaporator and a Kaufmann ion source for the IBAD process. The source material is MgO and a neutralized Ar^+ beam has an ion energy of 750 – 1000 eV oriented at 45° with respect to the substrate. Film growth is monitored by means of reflection high-energy electron diffraction (RHEED). The system allows for many experiments to be conducted in one pump-down due to the long length of metal substrate tape that is available. Experiments can be conducted either with moving tape or in segments with stationary tape. The system is ideal for high-throughput experimentation by linear combinatorial experimental design [5]. Hundreds of experiments can be spooled onto a reel of tape. The system is also utilized for fabrication of long length templates that are made for epitaxial deposition of high-temperature superconductors, such as $YBa_2Cu_3O_7$ (YBCO). This form of YBCO material is used for long length superconducting wire [6].

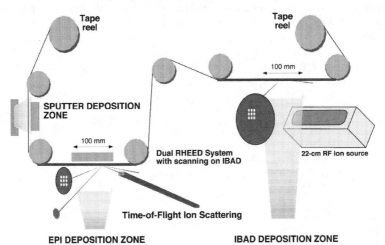

Fig. 3. Schematic of the IBAD deposition system at LANL. E-beam evaporation, as discussed in the text, is performed in the IBAD and Epi deposition zones. Tape can move in either direction and the deposited surface does not come in contact with the idlers.

Electropolished Hastelloy C-276 metal alloy is used as a substrate. The RMS surface roughness is typically less than 1 nm on a 5 x 5 μm area. The substrate is first ion-beam etched for one minute to clean off contaminants. A 10 nm thick amorphous Y_2O_3 is then deposited as a nucleation layer for IBAD-MgO. Following that IBAD of MgO is done at an MgO deposition rate of 0.1- 0.3 nm/sec and with an ion beam voltage of 1000 V and ion beam current of 50 – 100 mA (22 cm ion source). The ion-to-molecule ratio is varied between 0.5 and 1.0. The sample is approximately 35 cm from the ion source and 65 cm from the evaporation source. Quartz-crystal microbalances are used to monitor and control the deposition rate. RHEED images show biaxial crystalline alignment of the MgO from its early stages. The RHEED patterns are 3D spots and indicate an island morphology for the MgO film. Following the deposition of the IBAD layer, a thicker epitaxial layer is deposited on all the samples at a temperature of approximately 600°C. For the results shown here the epitaxial layer thickness is either 250 nm or 2 μm.

DISCUSSION

As has been reported earlier for IBAD-MgO texturing, the ion-beam induced texture evolves with thickness of the deposited layer [4,7]. Figure 4 shows the result of x-ray diffraction (XRD) analysis of samples with different IBAD layer thicknesses. While the XRD analysis shown is measured on the 250 nm thick homoepitaxial MgO layer, the results in the figure are due to the thickness variation of the IBAD layer. Pole figures are measured by XRD and the mosaic spread is characterized by the full width half maxima (FWHM) in ϕ and ω. We see that the FWHM in ϕ starts out at $\leq 15°$ and then decreases with increasing thickness to about 4°. However, the XRD intensity of the poles increases, reaches a maximum and then starts to decrease even before the optimum texture is achieved. At some point as the thickness is increased further, the XRD peaks lose all intensity as the crystalline-aligned fraction becomes insignificant.

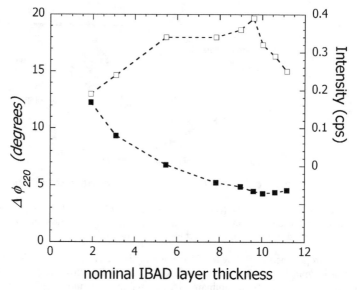

Fig. 4. Full-width half maxima for the (220) MgO peak in $\Delta\phi$ as a function of the IBAD layer thickness. Upper curve shows the XRD intensity of the peaks also as a function of thickness.

The texture of the MgO improves further with growth of thicker layers [8]. These layers can either be homoepitaxial MgO or heteroepitaxial layers, such as YBCO or Si. Eventually, this improvement is expected to stop, but we have not seen the FWHM saturate up to 2-3 μm. Figure 5 shows an XRD ϕ-scan for an MgO film that is 2 μm thick. In this case the in-plane texture was measured to be 1.9° and the out-of-plane texture was 1.0°. We have observed similar results for other heteroepitaxial layers on top of IBAD-MgO.

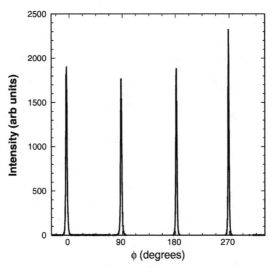

Fig. 5. Phi-scan for the MgO film with 1.9° in-plane FWHM. The MgO layer is 2 micrometers thick and is deposited on a polycrystalline metal alloy.

CONCLUSIONS

We have demonstrated that MgO can be textured biaxially using IBAD and the layer thickness can be optimized for minimum mosaic spread. For a thick homoepitaxial layer on top we demonstrated in-plane FWHM of 1.9° and an out-of plane FWHM of 1.0°.

ACKNOWLEDGMENTS

The authors would like to thank P. Arendt, J.R. Groves, R. DePaula, L. Stan, S. Kreiskott, and J. Haenisch for discussions. This work is supported by the Department of Energy Office of Electricity Delivery & Energy Reliability.

REFERENCES

1. J.M.E. Harper et al, Nucl. Instr. Meth. Phys. Res. B7/8, 886 (1985).
2. L.S. Yu, J.M.E. Harper, J.J. Cuomo and D.A. Smith, Appl. Phys. Lett. 47, 932 (1985).
3. Y. Iijima et al, Appl. Phys. Lett. 60, 769 (1992).
4. C.P. Wang et al, Appl. Phys. Lett. 71, 2955 (1997).
5. V. Matias and B.J. Gibbons, Rev. Sci. Instr. to be published (2007).
6. P.N. Arendt, S.R. Foltyn, Mater. Res. Soc. Bulletin 29, 543(2004).
7. A.T. Findikoglu et al, J. Mater. Res. 19 501 (2004).
8. V. Matias et al, IEEE Trans. Appl. Supercon. to be published (2007).

Mater. Res. Soc. Symp. Proc. Vol. 1020 © 2007 Materials Research Society   1020-GG02-06

Swift Heavy Ion Beam-Based Nanopatterning Using Self-Assembled Masks

Jens Jensen[1], Ruy Sanz[2], Marek Skupinski[1], Manuel Hernandez-Vélez[3], Göran Possnert[1], and Klas Hjort[1]

[1]Department of Engineering Sciences, Uppsala University, The Ångström Laboratory, Box 534, Uppsala, SE-751 21, Sweden

[2]Instituto de Ciencia de Materiales de Madrid, Consejo Superior de Investigaciones Cientificas, Madrid, 28049, Spain

[3]Departamento de Fisica Aplicada, Universidad Autonoma de Madrid, Madrid, 28049, Spain

ABSTRACT

Swift heavy ion beam-based lithography using masks of self-assembled materials has been applied for transferring well-ordered micro- and nanopatterns to rutile TiO_2 single crystals. As the induced damage has a high etching selectivity the patterns can be developed in HF with very high contrast. Here we present resulting patterns when using a mask of self-ordered silica spheres. Since the obtained structures are replicas of the mass distribution of the applied mask, the shape and size of resulting structures depend on the geometric configuration of the silica sphere layers. In addition, the resulting pattern can be tuned by varying the applied ion energy and fluence. Direct modifications of the optical properties of TiO_2 in a well-defined pattern are also presented.

INTRODUCTION

Fabricating regular micro- and nanostructures is of great interest due to their potential application in *e.g.* photonic crystals, data storage, displays, and biological sensors [1,2]. In projection lithography the pattern is defined by a mask, which provides the required template for producing the desired two- or three-dimensional arrays of structures.

Swift heavy ion irradiation induces localized material transformation in matter, so-called ion tracks [3], which, compared to *e.g.* e-beam or photo-lithography, gives a very high contrast between irradiated and non-irradiated regions. In addition, swift heavy ions can transform materials that are otherwise insensitive to electron or photon irradiation, they can induce high aspect ratio structures, and they are only minutely scattered. Combining swift heavy ions with high resolution absorbing masks may thus have potential as a lithography technique for nanotechnology. The advantage with this technique is that it enables, without further processing, directly patterned material modifications. Furthermore, creation of an ordered array of irradiation spots may offer advantages over *e.g.* the Focused Ion Beam technique, since masking would provide a parallel implantation and patterning technique. The requirements on the mask are very strict, however. The mask should be thick enough to stop or slow down the ions sufficiently and durable enough to withstand high-fluence irradiation or implantation without breaking, disintegrating, or swelling.

An interesting option as a lithographic mask is self-assembled materials with high-density nano- or micro-scale features. One example of a self-assembled template is the porous anodic alumina membrane (PAM) [1,2], see Figure 1(a). Only a few studies on pattern transfer using ion track lithography have until now been undertaken with this type of mask [4,5,6]. We have demonstrated pattern transfer to rutile TiO$_2$ single crystals resulting in nanopores with the same hexagonal ordering, having a diameter of 77 nm, 100 nm inter-pore distances, and 1100 nm depth [5], see Figure 1(b).

Another kind of promising mask consists of micro- to nanometre size colloidal particles [1,2]. With this kind of mask a hexagonal close-packed monolayer or multilayers of colloidal particles is formed on a surface and the open interstices between the particles can then be used as mask openings in lithography [7,8], Figure 1(c). There have only been a few investigations on ion beam-induced modification of the underlying substrate after high-energy irradiation through a single colloidal layer [9,10]. By swift heavy ion irradiation through a colloidal layer of silica spheres with a diameter of 1.6 μm, we were able to sculpture a TiO$_2$ surface into analogue patterns [10], see Figure 1(d).

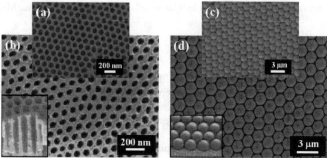

Figure 1. The synergetic approach used in micro/nanopatterning by ion irradiation. Masks produced by self-assembly (top pictures) and the resulting hexagonal patterns (bottom pictures) after lithography transfer. (a) Porous anodic alumina membrane and (b) resulting replicated nanopore pattern in TiO$_2$; (c) one monolayer of self-assembled silica spheres with ∅=1.6 μm, and (d) resulting surface on TiO$_2$. The inserts in (b) and (d) shows samples tilted by 30°. In both cases the samples were irradiated with 25 MeV Br ions using a fluence of 1×10^{14} ions/cm^2 prior to etching in HF.

Here we present further results on pattern transfer into rutile TiO$_2$ single crystals after MeV ion irradiation through self-assembled colloidal masks of silica spheres. TiO$_2$ has a high dielectric constant and is a semiconductor with a wide bandgap; both suitable characteristics in electronic and optical applications [11,12]. Furthermore, large surface areas and highly active sites of this material are useful in biotechnology and photocatalytic applications [11,13]. As ion tracks in TiO$_2$ have a very high etching selectivity, the induced damage can be developed in HF with very high contrast [14,15]. This makes is possible to prepare well-ordered arrays or patterns of TiO$_2$ with *e.g.* variation in refractive index or large surface areas.

EXPERIMENTAL

The experimental method is described in detail elsewhere [10]. Briefly, rutile single crystals TiO$_2$ (Crystal Gmbh) with (100) direction were used as substrates. Samples were exposed to UV irradiation to make the surface hydrophilic, thereafter a droplet of a colloidal suspension was deposited on the TiO$_2$ surface by means of a micropipette. The colloidal suspensions contained water and silica spheres with a mean diameter of 1.57±0.06 μm and 490±30 nm, respectively (Duke Scientific Corporation). The samples were then left to dry until all water had evaporated. To enhance the self-assembling and aggregation process of the silica spheres into a hexagonal close-packed structure, the samples were slightly tilted, making the evaporation process start from the upper parts of the samples [16]. In this manner, a monolayer of ordered silica spheres can be deposited on the surface.

Irradiations of the samples were done at the Tandem Laboratory, Uppsala University. The irradiations were performed at room temperature under normal incidence with respect to the sample surface using ion fluencies in the range $1.0x10^{12}$ to $1.0x10^{14}$ ions/cm^2. The energies and ion type used in this study were 8.5, 13 and 25 MeV Br ions. SRIM-2006 simulations [17] indicate that the ions completely pass through the silica spheres. Pristine rutile TiO$_2$ single crystal is not etchable by HF. Earlier ion track studies on rutile TiO$_2$ suggest that the electronic stopping power threshold for selective chemical etching of irradiated areas is $\approx$6.2 keV/nm [14,15], corresponding to $\approx$11 MeV Br ions. These studies indicated that it is the amorphous rutile phases and the stressed lattice region created by ions with energies above the threshold, which are highly soluble in HF. No etching is observed below the threshold value.

After irradiation the samples were immersed in ethanol and placed in an ultrasound bath to remove the silica spheres. Thereafter the samples were cleaned in a 1:1 H$_2$SO$_4$:H$_2$O$_2$ solution and rinsed with deionized water. Subsequently the samples were etched in a 20 % hydrofluoric acid (HF) solution for 35 minutes. The surfaces of the samples were investigated before and after etching by a LEO 1550 FEG high resolution scanning electron microscope (SEM).

RESULTS AND DISCUSSION

Large areas (up to a few mm^2) of ordered structures were obtained. Figure 1(d) shows the observed pattern on the surface after etching a TiO$_2$ sample irradiated through one monolayer of hexagonally ordered silica spheres with 25 MeV Br ions at a fluence of $1.0x10^{14}$ ions/cm^2. When seen from above, the diameter of the observed round structures is similar to that of the silica spheres. When tilting the sample in the microscope (see insert), the structures are seen to be hemispheres of TiO$_2$ lying on the surface with a height of $\approx$750 nm [10]. This is explained by the projected mass distribution of a silica sphere, which resembles a hemisphere. All ions going through a monolayer of silica spheres are able to deposit enough energy in TiO$_2$ to induce etchable damage. Nevertheless, the deposited energy gradually decreases when going outward from the centre of the silica spheres.

The etchable depth which can be reached in TiO$_2$ not only depends on the ion energy, but also on the applied ion fluence [5,10,14,15], as the etching properties depend on how large a fraction of the irradiated area is covered by induced amorphous and stressed lattice regions. For an applied fluence of $1.0x10^{14}$ ions/cm^2 a saturation value in etchable depth is reached [5,10,14,15]. This fluence dependence will influence the transferred pattern as can be seen in Figure 2(a-b) where TiO$_2$ substrates were irradiated with 13 MeV Br ions through one monolayer of silica spheres using ion fluencies of $4x10^{13}$ and $1x10^{14}$ ions/cm^2, respectively. In Figure 2(a) a visible, etched surface pattern is created. However, it was not possible to etch it into a three dimensional pattern. In Figure 2(b) the pattern of microspheres was more successfully transferred to the TiO$_2$. However, the obtained structures resemble truncated hemispheres, with a diameter similar to the silica spheres. This can be explained by the fact that only ions going through the peripheral parts of the silica spheres deposit enough energy in TiO$_2$ to induce etchable damage. The height between the non-etched plateau and the bottom is $\approx$700 nm.

Nevertheless, when applying silica spheres with a smaller diameter, 500 nm, ions going through the silica spheres can deposit enough energy in TiO$_2$ over the whole projected sphere area to induce etchable damage, even at 13 MeV. The transferred pattern is similar to Figure 1(d), but the structures are smaller, with a diameter similar to the silica spheres and a height of $\approx$300 nm. The surface of the TiO_2 hemispheres has a rough ('hairy') appearance after etching. This non-uniform etching may be attributed to fluctuation of the deposited energy in TiO_2 due to the statistical nature of the energy deposition [18]. Close to the chemical etching threshold, these density fluctuations may be rather large, reflecting a large dispersion in track diameter [18]. A higher energy seems to lower the observed roughness.

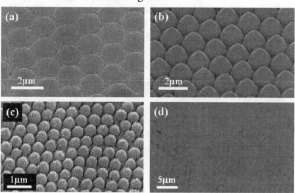

Figure 2. (a-c) Scanning electron micrographs of HF etched TiO_2 surfaces following irradiation with 13 MeV Br ions through one layer of silica spheres. Samples were tilted by ~30°. (a) Using an ion fluence of 4×10^{13} ions/cm^2 and spheres with $\varnothing$=1.6 µm. (b) Using an ion fluence of 1×10^{14} ions/cm^2 and spheres with $\varnothing$=1.6 µm. (c) Using an ion fluence of 1×10^{14} ions/cm^2 and spheres with $\varnothing$=500 nm. (d) Image taken in an optical microscope of a sample irradiated with 13 MeV Br ions at a fluence of 1×10^{14} ions/cm^2 through one layer of silica spheres with $\varnothing$=1.6 µm. The colloidal mask was removed before observation, but no etching was performed. A regular variation in reflectivity is seen.

Interestingly, a patterned material modification in TiO_2 was seen even without etching. Figure 2(d) shows a sample irradiated with 13 MeV Br ions at a fluence of 1×10^{14} ions/cm^2 through one layer of silica spheres with a diameter of 1.6 µm. After removing the mask the sample was observed in an optical microscope with the light applied from above. Light regions correspond to areas that reflect the light more strongly than darker area. A hexagonal pattern of circular regions of higher reflectivity is seen. This pattern is attributed to the masking effect of the silica spheres during irradiation. More material is affected by the irradiation between the spheres and the peripheral parts, inducing a modulation of the light scattering properties or refractive index of TiO_2 yielding the observed contrast. The regular variations in optical properties were seen even at fluencies where no etching was perceptible *i.e.* below 1×10^{13} ions/cm^2. It was also seen when using ion energies below the etching threshold.

Before removing the silica spheres from the irradiated TiO_2 surfaces, they were investigated by SEM to check whether they were subjected to any deformation. Under exposure to MeV ions, colloidal particles may undergo extreme deformations induced by the deposited energy [8]. Deformation of the silica spheres has been observed to be energy, temperature and fluence dependent [8], and the modification of their shape will influence the transferred pattern. However, for the used maximum fluence of 1.0×10^{14} ions/cm^2 we only saw minute changes in shape of the silica spheres for all energies.

What makes silica spheres interesting as a mask material is the fact that multilayers can be created, and therefore different patterns can be obtained. Figure 3(a) shows an area which has been irradiated with ions going through a mask consisting of two hexagonally ordered layers. Spheres in the second layer are placed above alternate interstices in the first layer. After etching, the TiO_2 surface looks like a hexagonally ordered array of holes having diameters of ≈300 nm. The depth of these holes is ≈650 nm. The surface has a bumpy appearance. Decreasing the sphere diameter to 500 nm, Figure 3(b), the pattern resembles flowers with six petals, and the features have a three dimensional characters. As in Figure 1(c), the surface is quite rough. The holes are approximately 150 nm. This pattern was also observed after irradiating through 1.6 μm silica spheres with 25 MeV, however with larger (and clearer) features and less rough surface [10]. The protruding character of the petals gives the surface in Figure 3(a) a bumpy appearance.

Figure 3. Scanning electron micrographs of HF etched TiO_2 samples following irradiation with Br ions at a fluence of 1×10^{14} ions/cm^2 through different colloidal masks. (a) Mask consisted of two layers of hexagonally ordered silica spheres with ∅=1.6 μm, and ion energy 13 MeV. Sample was tilted by ~30° (Note the 'defect' due to missing silica spheres in the mask). (b) Use of mask of two layers of hexagonally ordered silica spheres with ∅=500 nm, and ion energy 13 MeV. Sample was tilted by ~30°. (c) Use of mask consisting of three layers of hexagonally ordered silica spheres with ∅=1.6 μm, and ion energy 25 MeV.

When the number of layers is further increased, it becomes harder to induce etchable damage yielding an appreciable pattern, using 13 MeV. Higher energies are necessary. Figure 3(c) shows the result of irradiating with 25 MeV ions through a mask of three layers of hexagonally ordered silica spheres. The spheres in the third layer are placed directly above those interstices in the first layer that were not covered with spheres in the second layer. The resulting array in TiO_2 has holes with a diameter of ≈300 nm and a depth of ≈1 μm. The patterns in Figure 3(a) and (c) are very similar to the well-ordered pattern of a PAM we have transferred to TiO_2 using ion track lithography; see Figure 1(b). However, using high aspect ratio PAM as masks, mapping by Rutherford Backscattering Spectrometry is used to align the mask pores with the ion beam [4,5]. However, for irradiation through a self-assembled pattern of silica spheres no special mask alignment is needed.

Besides the hexagonally ordered layers we have also observed other configurations, *e.g.* quadrangular packaging, leading to interesting patterns [10]. In addition, the colloidal mask contains in some places defects like dislocations, a boundary between two mask configurations, and vacancies (see Figure 3(a)). The etched pattern after irradiation through a single monolayer is easily predicted. However, in order to better understand the etched structures, obtained when ions pass through several equivalent layers of spheres, we have made simple simulations [10].

CONCLUSIONS AND OUTLOOK

This study shows that swift heavy ion irradiation through a mask of self-assembled material enables fabrication of regular micro- and nano-structures in rutile TiO_2 single crystals. As the induced damage has a very high etching selectivity, the patterned damage can be developed in HF with high resolution. Using a mask of self-organized silica spheres, the transferred pattern depends on the geometrical configuration of the mask. The possibility of tuning the colloidal particles in a desired pattern will be further studied. Furthermore, this ion beam-based lithography technique allows a direct modification of the optical properties of TiO_2 in a well-defined pattern. A study of the physical nature of the optical changes, and its dependence on ion energy and fluence are in progress.

ACKNOWLEDGMENTS

The authors would like to thank the staff at the Tandem Laboratory, Uppsala University, for technical assistance. Jens Jensen thanks the Carl Tryggers Foundation for financial support.

REFERENCES

1. H.S. Nalwa, *Handbook of Nanostructures Materials and Nanotechnology*, (Elsevier, Amsterdam 1999).
2. B. Bhushan (editor) *Handbook of Nanotechnology*, (Springer Verlag 2004).
3. R. Spohr, *Ion Tracks and Microtechnology, Principles and Applications*, (Vieweg & Sohn Verlagsgesellschaft mbH, Braunschweig, 1990).
4. A. Razpet, A. Johansson, G. Possnert, M. Skupinski, K. Hjort, A. Hallén, J. Appl. Phys. **97**, 44310 (2005).
5. R. Sanz, A. Johansson, M. Skupinski, J. Jensen, G. Possnert, M. Boman, M. Vazquez, K. Hjort, Nano Letter **6**, 1065 (2006).
6. M. Skupinski, J. Jensen, A. Johansson, A. Razpet, G. Possnert, M. Boman, K. Hjort, submitted to J. Vac. Science and Tech. B (2006).
7. F. Burmeister, W. Badowsky, T. Braun, S. Wieprich, J. Boneberg, P. Leiderer, Appl. Surf. Sci. **144/145**, 461 (1999).
8. T. van Dillen, A. van Blaaderen, A. Polman, Mater. Today **7/8**, 40 (2004), and references therein.
9. C. Strohhöfer, J.P. Hoogenboom, A. van Blaaderen, A. Polman, Adv. Mater. **14**, 1815 (2002).
10. M. Skupinski, R. Sanz, J. Jensen, Nucl. Instrum. Meth. B, in press (2007).
11. U. Diebold, Surf. Sci. Rep. **48**, 53-229 (2003).
12. K. Rajeshwar, N.R. de Tacconi, C. R. Chenthamarakshan, Chem. Mater. **13**, 2765 (2001).
13. G.K. Mor, K. Shankar, M. Paulose, O.K. Varghese, C.A. Grimes, Nano Lett. **5**, 191 (2005).
14. K. Nomura, T. Nakanishi, Y. Nagasawa, Y. Ohki, K. Awazu, M. Fujimaki, N. Kobayashi, S. Ishii, K. Shima, Phys. Rev. B **68**, 064106 (2003).
15. K. Awazu, M. Fujimaki, Y. Ohki, T. Komatsubara, Radiation Measurements **40**, 722 (2005).
16. R. Micheletto, H. Fukuda, M. Ohtsu, Langmuir **11**, 3333 (1995).
17. www.srim.org.
18. J. Jensen, A. Razpet, M. Skupinski, G. Possnert, Nucl. Instrum. Meth. B **243**, 119 (2006), and references therein.

Mater. Res. Soc. Symp. Proc. Vol. 1020 © 2007 Materials Research Society   1020-GG02-09

High-Aspect-Ratio Micromachining of Fluoropolymers Using Focused Ion Beam

Yoshinori Matsui[1], Nozomi Miyoshi[2], Akihiro Oshima[2], Shu Seki[1], Masakazu Washio[2], and Seiichi Tagawa[1]

[1]The Institute of Scientific and Industrial Research, Osaka University, 8-1 Mihogaoka, Ibaraki, Osaka, 567-0047, Japan

[2]Research Institute for Science and Engineering, Waseda University, 3-4-1 Okubo Shinjuku-ku, Tokyo, 169-8555, Japan

ABSTRACT

Poly(tetrafluoroethylene) (PTFE) microstructure with high aspect ratio (> 200) and without solid debris along the edge was fabricated with high etch rate using FIB. Evolution of PTFE by FIB is responsible for the high aspect ratio, the high etch rate, and the no solid debris. Roughness of etched surface of the PTFE increases with fluence, although edge of the etched area has good profiles. The etch mechanism seems to be complicated.

INTRODUCTION

Fluoropolymers have excellent properties such as high thermal stability, high chemical stability, low adhesion, biological suitability, and low frictional resistance [1]. Micromachining of these polymers are very attractive for biomedical microelectromechanical system (BioMEMS) applications, e.g., flow cytometers, biological assays, and nano-filters. However, hand fabrication limits the ability of current devices to be effectively reduced in size. There was no suitable solvent for chemical etching of these polymers to perform wet bulk micromachining, thus micromachining of these polymers had been considerably difficult. Micromachining of these polymers have been intensively studied by pulsed laser ablation [2-4], or photo-etching with synchrotron radiation [5-7]. Micromachining with laser ablation is however not preferred, because resolution is poor and aspect-ratio, which is the ratio of the depth of microstructures to the pattern width, is quite low (~2). Photo-etching by synchrotron radiation enable fluoropolymers to be etched with high-aspect-ratio [5-7]. This method uses contact mask with high cost, and the resolution partially depends on the mask with thickness enough to cut off the radiation. In order to realize applications in BioMEMS, micromachining of fluoropolymers with high resolution, high aspect ratio, high etching rate, excellent etch profiles is desirable.

Focused ion beam micromachining (FIBM) is appropriate to achieve nanometer - micrometer modification because of the potential for allowing high resolution (approaching tens of nanometer). FIBM is carried out by scanning the ion beam over the desired area with material removal via radiation chemical reaction and physical spattering. Unfortunately, the physically sputtered materials redeposit along the edge of the etched area, and at last the sputtered materials cannot be removed from etching area. This process results in low etching rate, etch profiles with low aspect ratio, and reducing the resolution. Redepositon of removed material is therefore one of the most serious problem in FIB fabrication.

In the present study, we carried out micromachining of PTFE by FIB to produce microstructures and investigated outgassed species from irradiated fluoropolymers using in-situ quadrupole mass spectrometry to make clear etching mechanism of PTFE by FIB.

EXPERIMENT

Irradiation of FIB was carried out by using Seiko Instruments Inc. SMI2050 (Figure 1). The FIB equipped with Ga ion, and acceleration voltage set at 30 kV. Beam current and diameter of the FIB are controlled by the metal film, which has apertures with various diameters, inserted between condenser and objective lenses. The beam currents (and beam size (fwhm)) employed were 1.3 nA (100 nm), 2.9 nA (200 nm), 5.3 nA (300 nm). FIB was exposed into spots at intervals of the beam size (fwhm), and the exposure time at each spot was 100 μs.

In order to examine outgassed species from irradiated samples, sample chamber of the FIB system is connected with a PFEIRRER VACUUM Prisma™ QMS 200 M3 mass spectrometer with a 100 mA, 300 V ionizing source. The QMS was used in scan mode which was used to determine the masses of all outgassed fragments by scanning over 1-300 amu. The sample chamber was evacuated to a back ground pressure about 5×10^{-7} torr. The total gas pressure was measured by a PFEIRRER VACUUM full rage vacuum gauge. Details of the system have been described elsewhere [8].

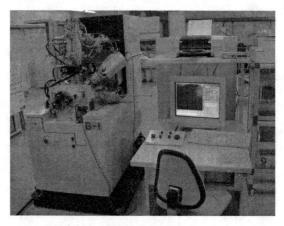

Figure 1. FIB system connected with QMS.

PTFE sheets with a thickness of 50 - 5000 μm were purchased from NICHIAS Corporation. The PTFE sheets were cleaned with methanol prior to use. Microstructures made were observed by a JEOL JSM-6335 field emission scanning electron microscope.

DISCUSSION

Figure 2a shows etch profiles of PTFE which was exposed to FIB with a beam current of 1.3 nA at various fluences. Etch depth increases with fluence in ranging of $1 \times 10^{16} \sim 6.4 \times 10^{17}$ ions/cm^2. In contrast with the case in poly(methylmethacrylate) (PMMA) and polystyrene (PS) [9], the edges of these etch profiles is defined sharply and no solid debris is found at the edge. Except for etched PTFE surface, the quality of the microstructure made by the FIB is quite good, which has been likely seen in PTFE microstructure made by SR [5].

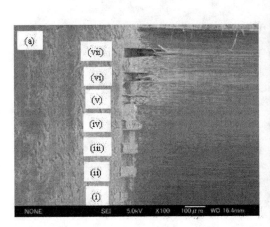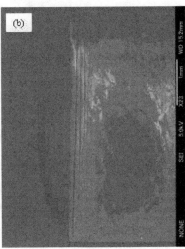

Figure 2. (a)SEM image of the microstructure in a PTFE sheet by FIB with the beam current of 1.3 nA under the fluence of (i) 1×10^{16}, (ii) 2×10^{16}, (iii) 4×10^{16}, (iv) 8×10^{16}, (v) 1.6×10^{17}, (vi) 3.2×10^{17}, and (vii) 6.4×10^{17} ions/cm^2. (b) Cross sectional view of etched line in 5 mm thick PTFE sheet with the beam current of 1.3 nA.

Figure 3 shows morphological change of the PTFE surface etched by FIB with a beam current of 1.3 nA at various fluences. Roughness of the PTFE increases when exposed to FIB, and then network structures and projections perpendicular to the surface are formed as shown in Figure 3b. As the PTFE surface is etched, the length of the projections increases and the projections seemed to get together. For instance, a projection in Figure 3d seemed to be built up of many fine projections which were likely seen in Figure 3c. Etch mechanism of PTFE by FIB seems to be complicated. For instance, some polymers undergo closslinking reaction as well as degradation reaction by ion beams [10,11]. Figure 4 shows fluence dependence of etch depth and roughness of etched PTFE surfaces. Etch depth and the roughness simultaneously increase with fluence. Width of the etched surface becomes gradually narrower as PTFE was etched. The cross-sectional view of etched line is shown in Figure 2b. It can be seen from the figure that aspect ratio of the FIBM is surprisingly more than 200.

Table 1 summarizes etch rate of PTFE sheet in the range of the lower fluence. The etch rate with the beam current of 1.3 nA was 8.0×10^{-20} g/ion, indicating surprisingly that 480 of CF_2CF_2 unit was removed by single ion. The rate is several hundred times higher than that of PS in the range of the lower fluence (Etch rate of the PMMA and PS sheet become low as the PS sheet is etched because of the solid debris at the bottom [9]). The etch rate of PTFE increases with beam

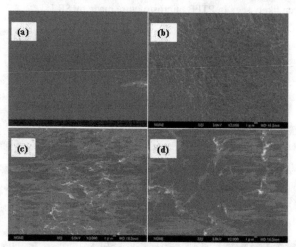

Figure 3. SEM image of bottom of etched area in PTFE sheet irradiated with FIB with the beam current of 1.3 nA under the fluence of (a) 0, (b) 1×10^{16}, (c) 4×10^{16}, and (d) 1.6×10^{17} ions/cm^2.

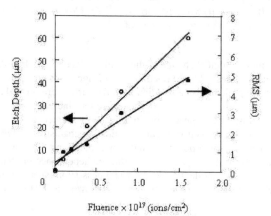

Figure 4. The etch depth and roughness of bottom of etched area in PTFE sheet irradiated with FIB with the beam current of 1.3 nA.

current. FIB-induced local heating in PTFE seemed to play an important role for etch processes.

Figure 5 shows mass spectra of outgassed species from PTFE FIB irradiated by FIB. The mass spectra of outgassed species were obtained by subtraction of the background spectra from the ones observed during irradiation. In contrast to the mass spectra of outgassed species from irradiated PS (no significant peaks in the region more than 200 amu [9]), there are many mass spectral peaks in the region more than 200 amu. The peak located at 69 amu is assigned to CF_3.

Table 1. Etch rate for PTFE fabricated by Ga^+ FIB with various beam currents.

Beam Current	Etch Rate (μm/s)	Etch Rate (g/ion)* $\times 10^{-19}$	Number of the Removed CF_2CF_2
1.3	0.033	0.80	480
2.9	0.11	1.2	690
5.3	0.21	1.3	750

Etched Area is 50 μm × 30 μm. Density of PTFE is 2.14 g/cm³.

Table 2. Molecular weights and boiling points of fluorocarbons

	Molecular Weight	Boiling Point (°C)
C_3F_8	188	-36.6
C_4F_{10}	238	-1.9
C_5F_{12}	288	29.2
C_6F_{14}	338	56.6

Source: Lide, D. R. Handbook of Chemistry and Physics 81st Edition, CRC Press, Boca Raton, London New York, Washington, D. C., 2000-2001

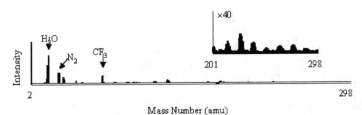

Figure 5. Mass spectra of outgassed species from PTFE sheet irradiated by FIB with the beam current of 1.3 nA.

Other significant peaks are due to fragments of C_xF_y. Besides, it is predicted that outgassed species with mass-number greater than 300 amu evolve from irradiated PTFE. Table 2 shows boiling point of hydrocarbons and fluorocarbons. It can be seen from Table 2 that the boiling point of fluorocarbons were quite low with compared to hydrocarbons. Outgassed flurorocarbons with high molecular weight are produced by main chain scission of PTFE, and evolve from irradiated PTFE as gaseous products because of their low boiling point. Therefore, there are no solid debris in the PTFE microstructure, leading to high etch rate and high aspect ratio. G-value, which means the number of specified chemical events produced in an irradiated substance per 100 eV of energy absorbed from ionizing radiation, is 3.2 for PTFE micromachining by FIB with 1.3 nA if we assume molecular weight of fluorocabons removed was 300. This indicate that PTFE is effectively etched by FIB.

CONCLUSIONS

We carried out FIB micromachining of PTFE, and found that microstructures of PTFE can have high aspect ratio (more than 200) and no debris along the etched area. The etch rate with the beam current of 1.3 nA was 8.0×10^{-20} g/ion, indicating that 480 of CF_2CF_2 unit was surprisingly removed by single ion. Outgassed species from irradiated PTFE were investigated by *in-situ* quadrupole mass spectrometer to make clear etch mechanism. As for the mass spectroscopic study, outgassed fluorocarbons with high molecular weight (69 ~ 298) and low boiling point evolve from PTFE irradiated by FIB. Therefore, PTFE microstructure was made with high etch rate, high aspect ratio, and no debris along the microstructure. Micromachining of PTFE has an excellent potential in making microparts for BioMEMS applications.

REFERENCES

1. Mark, H.; Bikales, N.; Overberger, C.; Menges, G. in: *"Encyclopedia of Polymer Science and Engineering"*, 2nd ed., J. Wiley & Sons, New York, **1964**, Vol. 16, p. 577.
2. Küper, S.; Stuke, M. *Appl. Phys. Lett.* **1989**, 54, 4.
3. Kumagai, H.; Okamoto, T.; Obara, M. *Appl. Phys. Lett.* **1994**, 65, 1850.
4. Riedel, D.; Castex, M. C. *Appl. Phys.* **1999**, A 69, 375.
5. Zhang, Y.; Katoh, T.; Washio, M.; Yamada, H.; Hamada, S. *Appl. Phys. Lett.* **1995**, 67, 872.
6. Katoh, T.; Nishi, N.; Fukagawa, M.; Ueno, H.; Sugiyama, S. *Sensors and Actuators*, 2001, A89, 10.
7. Yamaguchi, D.; Katoh, T.; Sato, Y.; Ikeda, S.; Hirose, M.; Aoki, Y.; Iida, M.; Oshima, A.; Tabata, Y.; Washio, M. *Macromol. Symp.* 2002, 181, 201.
8. Matsui, Y.; Seki, S.; Tagawa, S. *J. Photopolym. Sci. Technol.* **2005**, 18, 501.
9. unpublished data
10. Seki, S.; Maeda, K.; Tagawa, S.; Kudoh, H.; Sugimoto, M.; Morita, Y.; Shibata, H. *Adv. Mater.*, **2001**, 13, 1663.
11. Seki, S.; Tsukuda, S.; Maeda, K.; Matsui, Y.; Saeki, A.; Tagawa, S. *Phys. Rev. B*, **2004**, 70, 144203.

Mater. Res. Soc. Symp. Proc. Vol. 1020 © 2007 Materials Research Society   1020-GG03-04

Resolution Performance of Programmable Proximity Aperture MeV Ion Beam Lithography System

Sergey Gorelick[1], Timo Sajavaara[1], Mikko Laitinen[1], Nitipon Puttaraksa[2], and Harry J. Whitlow[1]

[1]Dept. of Physics, University of Jyväskylä, Jyväskylä, 40014, Finland
[2]Dept. of Physics, Chiang Mai University, Chiang Mai, 50200 (FNRF), Thailand

ABSTRACT

An ion beam lithography system for light and heavy ions has been developed at the University of Jyväskylä's Accelerator Laboratory. The system employs a programmable proximity aperture to define the beam. The proximity aperture is made up of four Ta blades with precise straight edges that cut the beam in the horizontal and vertical directions. The blade positions and dimensions are controlled by a pair of high-precision linear-motion positioners. The sample is mounted on a X-Y-Z stage capable of moving with 100 nm precision steps under computer control. The resolution performance of the system is primarily governed by the proximity aperture. Pattern edge sharpness is set by the beam divergence, aperture blade straightness, and secondary and scattered particles from the aperture blade edges. Ray tracing simulations using object oriented toolkit GEANT4 were performed to investigate the beam spatial resolution on the sample defined by the proximity aperture. The results indicate that the edge-scattering does not significantly affect the pattern edge sharpness.

INTRODUCTION

MeV proton beam writing (PBW) is a rapidly evolving lithography technique capable of patterning high 3D nanopatterns with very high line-width to resist-thickness aspect ratio (more than 100), and with a high resolution of better than 20 nm [1-3]. The PBW technique is analog to electron beam lithography (EBL). A beam of protons from an electrostatic accelerator is magnetically focused and scanned over the resist surface. However, the MeV protons, as opposed to keV electrons, can penetrate deep into the resist along a straight path with minimal scattering.

Proton beams from cyclotrons generally have higher energies (tens of MeV), which enables pattern writing in thicker resists (up to 400 μm for 6 MeV protons in PMMA [4]). However, even if large beam currents are available (up to 100's of μA), the rather large divergence (about 1 mrad) and poor energy resolution imply that it is not straightforward to use a focusing in order to reach μm beam spot sizes. An alternative approach is to raster the target relative to a small beam spot defined by an aperture [5,6]. In MeV ion programmable proximity aperture lithography (PPAL), which is used in our system, a rectangular beam spot is defined by a "shadow" of a computer-controlled variable aperture in close proximity to the sample (Fig. 1). The aperture is made up of two L-shaped Ta blades with straight edges (each blade is made up of two 100 μm thick Ta sheets glued together). Precise movement of each L-shaped blade in the X' and Y' directions defines the size of the beam spot (Fig. 1(a)). By combining X' and Y' movement of the defining aperture with x and y movement of the target, entire rectangular pattern elements up to 500×500 μm² can be written in a single exposure over 20×20 mm² field. A technical overview of the PPAL system can be found elsewhere [7].

The PPAL approach considerably speeds up the writing time for large patterns compared to writing by scanning a small beam spot over the entire pattern element. This system is a valuable tool for our biomedical research programs at a cellular and sub-cellular level [8], where we are interested in rapidly exposing patterns with a large numbers of pattern elements of 10-300 μm size over a large area in thick (≤200 μm) resists in order to form cell-growth substrates. However, we believe that the PPAL system, besides production of cell culture substrates, may be also employed for a rapid production of 3D micro- and nanostamps, lab-on-chip and fluidic devices, micro- and nanophotonics structures.

The pattern edge sharpness is set by the beam divergence, aperture blade straightness, and secondary and scattered particles from the aperture edges. Ray tracing simulations were performed to investigate the beam spatial resolution on the sample as defined by the proximity aperture taking into account the beam divergence.

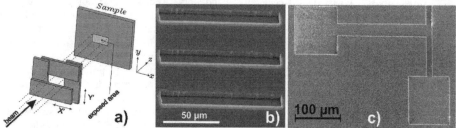

Figure 1. PPAL system. a) Two L-shaped aperture blades (each blade is made up of two 100 μm thick Ta sheets glued together). By combining the X' and Y' movement of the blades with x and y movement of the target entire rectangular pattern element of various sizes can be written in one step. b) and c) Examples of patterns written with PPAL system in 7.5 μm thick PMMA using 56 MeV ¹⁴N beam.

SIMULATION OF THE EXPERIMENTAL SET-UP

The GEANT4 toolkit [9] was used to test the concept of the PPAL system. GEANT4 is an object-oriented Monte Carlo simulation tool written in C++ that provides comprehensive detector and physics modeling capabilities. Promising results of simulations at micrometer scale [10,11] made GEANT4 a natural choice to investigate the PPAL system resolution performance. In GEANT4 users can define their own materials and system geometry. We have defined four 100 μm thick Ta sheets combined together in space to form the defining aperture. The primary Ta circular aperture (2 mm thick and 1 mm diameter) was placed 1.76 m upstream from the defining aperture. Collisions with the residual gasses in the beam line were neglected in order to investigate a pure slit-edge scattering. The PPAL system was investigated using the latest release of GEANT4 (version 4.8.2) and its low energy extension (G4LOWEM3.0). Ion processes, such as multiple scattering and ionization, were simulated using standard EM physics provided by GEANT4 [12]. The general particle source (GPS) was used to produce proton beams with the desired energy, angular and spatial distributions. To provide reliable multiple scattering of the protons in the primary aperture and in the defining aperture's blades the maximal simulation step was set to 50 nm. The value of the secondary particle cut has been set to 100 μm in the primary

Ta aperture and in the four Ta sheets for 2 MeV protons. The values for the cut-value and the maximal simulation step have been optimized from a comparison between GEANT4 and SRIM 2006 [4] beam straggling simulations. In this calibration 2 MeV protons entering perpendicularly to the target surface passed through Ta foils of different thicknesses, and the positions of exiting protons were measured. Having position distributions of the transmitted protons it is possible to calculate quantities characterizing scatterings in the material, such as lateral projected range, lateral straggling, radial range and radial straggling. In this study, we used definitions of these quantities proposed by SRIM 2006 [4]. The GEANT4 and SRIM 2006 predictions show a reasonable agreement over the range of 1-15 μm of Ta foil thicknesses (Fig. 2). This suggests we can be confident in the predictions obtained in our calculations.

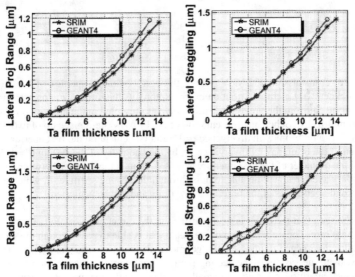

Figure 2. Comparison of 2 MeV proton beam straggling in Ta simulated by SRIM 2006 and GEANT4.

RESULTS AND DISCUSSION

To investigate the beam spot spatial resolution we modeled the target as a sensitive detector (sample) that could be placed at different distances behind the defining aperture. The detector recorded each proton's hit position coordinates (x,y) and its energy. This information could then be stored in files and processed by numerical analysis software, such as Matlab or ROOT [13]. In initial simulations we found out that the protons, that undergo scattering at the edges of the first aperture, have a negligible probability to transverse through the defining aperture. In these simulations the diverging (1 mrad) beam with a 600 μm radius was incident on the first collimator 1 mm in diameter. Fig. 3(a) shows the spatial distribution of protons transmitted through the first collimator and recorded 1 m downstream (0.76 m before the defining aperture). For 600 μm radius proton beam centered on 500 μm radius collimator with 3.4×10^6 incident protons we obtain 2.36×10^6 transmitted, giving 69.4% transmission efficiency in the simulation, which is the proton beam spot to collimator opening surface ratio ($\pi \cdot 500^2$/ $\pi \cdot 600^2$). Among the transmitted protons 0.29% or 6865 protons have been scattered by the collimator edges. Because the detector had a form of a 500×500 mm^2 square we estimate that the number of the scattered protons could be larger, since those hits further than 250 mm from the aperture center were not recorded.

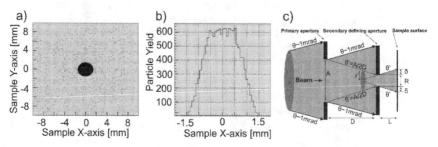

Figure 3. Transmission of 2 MeV protons through the first Ta aperture of 1 mm diameter and 2 mm thickness. a) Lateral distribution of the proton hits measured 1 m downstream b) 10 µm thick stripe cut from the previous distribution and presented in the histogram form, particle yield not normalized, c) Schematic presentation of the PPAL system made up of two apertures (not to scale). A proton beam with divergence θ=1 mrad is incident on the first aperture. Only that fraction of the beam having divergence set by the geometry of the system θ'=A/2D, where A=1 mm is the diameter of the first aperture and D=1.76 m is the distance between the primary and secondary apertures, will be able to pass through the second aperture. R, which is the opening of the second aperture, is of the order of a few µm's and can be neglected.

It can be seen that the beam is well-confined at the target position (Fig. 3(a-b)), while scattering events are rare. The only fraction of the beam that is able to pass through the second aperture has a smaller divergence defined by the geometrical dimensions of the system. Therefore, we can approximate the system by neglecting the primary aperture, moving the proton source closer to the second aperture and making it smaller, and setting the maximal divergence to θ'=A/2D=0.28 mrad instead of θ=1 mrad (see Fig. 3(c)). Such approximation allowed speeding up considerably (factor of 10) of the Monte Carlo calculations by allowing a smaller number of protons to be used. In our simulations we used a proton source of r'=50 µm radius with 0.28 mrad divergence placed 176 mm upstream the second aperture. Typically, 7.86×10^6 protons were simulated for aperture openings 1×10 µm^2 and 10×10 µm^2, while 26.5×10^6 protons for 0.1×10 µm^2 opening to ensure good statistics. The spatial distributions of the proton hits were measured at 0.5 mm, 1 mm, 2 mm and 5 mm distances behind the fourth Ta blade. For 7.86×10^6 incident protons, on average 9.5×10^4 and 9.5×10^3 are transmitted, and 2.6×10^4 and 1.8×10^3 are scattered through 10×10 µm^2 and 1×10 µm^2 apertures, respectively. For 26.5×10^6 incident protons, on average 3×10^3 are transmitted and 6×10^4 are scattered through 0.1×10 µm^2 aperture. Fig. 4(a) shows a typical spatial distribution of protons transmitted and scattered through the second aperture. The transmitted protons create rectangles well-confined within the intended locations (Fig. 4(b)). The scattered protons create a large "halo" spreading to several tens of mm. The protons are scattered mainly to the upper left corner of the sample (due to the geometry of the Ta blades). The scattered protons are separated by a few µm's and sometimes even by a few mm. The results indicate that the slit-edge scattering will not degrade the pattern edge sharpness. The pattern dimensions will therefore be mainly set by the beam divergence and the Ta blades sharpness.

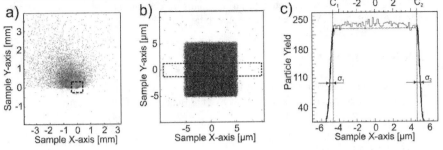

Figure 4. a) Spatial distribution and energy of the protons passing through $10\times10 \ \mu m^2$ second aperture measured 1 mm behind the fourth Ta blade. b) Distribution of the transmitted particles. c) Histogram built from 2 μm thickness stripe cut from the transmitted proton data, particle yield not normalized. FWHM of the beam spot can be found from Gaussian fits to the histogram tails, $FWHM=(C_2+1.7741\cdot\sigma_2)-(C_1-1.7741\cdot \sigma_1)$.

Stripes of 2 μm thickness from the hits distributions were "cut" and plotted in histogram form. The FWHM's of the beam spots were found from the Gaussian fits to the histograms' tails (e.g. see Fig. 4(c)). Fig. 5(a-c) show FWHM of the beam spots "shadowed" by the $10\times10 \ \mu m^2$, 1×10 μm^2, and $0.1\times10 \ \mu m^2$ aperture openings plotted against the distance between the sample and the fourth Ta blade. Linear functions were fitted to the results.

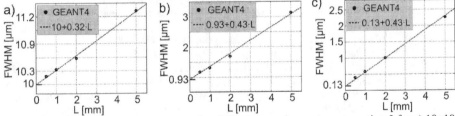

Figure 5. FWHM of beam spots measured at different sample-aperture separation L for a) 10×10 μm^2, b) $1\times10 \ \mu m^2$ and c) $0.1\times10 \ \mu m^2$ aperture openings.

From the linearity of the spot dimensions it follows that the beam divergence is responsible for the spot broadening, whereas contribution from the edge-scattering is negligible. Note, however, that because we used maximal allowed beam divergence, the calculated beam-spot spreading on targets presents an upper limit. Realistic beam spot spreading is anticipated to be smaller since accelerator beams are significantly paraxial.

CONCLUSIONS

Ray-tracing simulations using GEANT4 toolkit were performed to calculate the PPAL system's resolution performance. The calculations indicate that the slit edge-scattering will not degrade the pattern edge sharpness. The beam spot spreading is mainly associated with the beam divergence. Since maximal possible beam divergence, set by the geometry of the system, was used in this paper, the calculated beam spreading represents an upper limit. The resolution performance of the PPAL system is expected to be better due to a significant paraxial component of typical accelerator beams.

ACKNOWLEDGMENTS

SG is grateful for travel support from Vilho, Yrjö ja Kalle Väisälän rahasto foundation. The work was carried out under the auspices of the Academy of Finland Centre of Excellence in nuclear and accelerator-based physics (Ref 213503).

REFERENCES

1. F. Watt, A.A. Bettiol, J.A. van Kan, E.J. Teo, M.B.H. Breese, International Journal of Nanoscience **4**, 269 (2005).
2. J.A. van Kan, A.A. Bettiol, F. Watt, Nano Letters **6**, 579 (2006).
3. J.A. van Kan, P.G. Shao, K. Ansari, A.A. Bettiol, F. Watt, Microsystem Technologies **13**, 431 (2006).
4. J.F. Ziegler, SRIM-2006, from http://www.SRIM.org
5. J.V. Erps et. al., IEEE Photonics Technologiy Letters **18**, 1165 (2006).
6. M.L. Taylor, A. Alves, P. Reichart, R.D. Franich, S. Rubanov, P. Johnston, D.N. Jamieson, Nucl. Instr. and Meth. B (2007), doi:10.1016/j.nimb.2007.02.057.
7. S. Gorelick, T. Ylimäki, T. Sajavaara, M. Laitinen, A. Sagari A.R., H.J. Whitlow, Nucl. Instr. and Meth. B (2007), doi:10.1016/j.nimb.2007.01.260.
8. S. Gorelick, P. Rahkila, A. Sagari A.R., T. Sajavaara, S. Cheng, L.B. Karlsson, J.A. van Kan, H.J. Whitlow, Nucl. Instr. and Meth. B (2007), doi:10.1016/j.nimb.2007.02.008.
9. S. Agostinelli et. al., Nucl. Instr. and Meth. A 506, **250** (2003); from http://geant4.web.cern.ch/geant4/
10. S. Incerti, C. Habchi, Ph. Moretto, J. Olivier, H. Seznec, Nucl. Instr. and Meth B **249**, 738 (2006).
11. S. Incerti, Ph. Barbaret, B. Courtois, C. Michelet-Habchi, Ph. Moretto, Nucl. Instr. and Meth. B **210**, 92 (2003).
12. GEANT4 Physics Reference Manual, available at http://geant4.web.cern.ch/geant4/UserDocumentation/UsersGuides/PhysicsReferenceManual/html/index.html
13. ROOT – an object-oriented data analysis framework, from http://root.cern.ch

Patterning, Quantum Dot Synthesis, and Self Assembly

Mater. Res. Soc. Symp. Proc. Vol. 1020 © 2007 Materials Research Society   1020-GG05-01

Patterned Adhesion of Cells

Robert Lee Zimmerman[1], Ismet Gurhan[2], and Daryush ILA[1]

[1]Center for Irradiation of Materials, Alabama A&M University, 3900 Meridian Street, Normal, AL, 35762

[2]Faculty of Engineering, Ege University, Izmir, Turkey

ABSTRACT:

It is well known that silver deposition avoids bacterial growth and inhibits the natural process of attachment of connective tissue to biocompatible materials *in vivo*. We have completed a five year investigation of the precise spatial control of cell growth on glassy polymeric carbon implanted with silver using ion beam techniques, and the persistence of the inhibitory effect on cell growth. Long term inhibition of cell growth on GPC is a desirable improvement on current cardiac implants and other biocompatible materials placed in the blood stream. We have used implanted silver ions near the surface of GPC to completely inhibit cell attachment and adhesion. Cells attach and strongly adhere to areas close to the silver implanted surfaces. Patterned ion implantation permits precise control of tissue growth on GPC and other biocompatible substrates. Cell growth limited to micrometric patterns on a substrate may be useful for *in vitro* studies of associated biological processes in an otherwise identical environment. The patterned inhibition of cell attachment persists for periods of time significant relative to typical implant lifetimes.

INTRODUCTION:

Glassy Polymeric Carbon (GPC), like other forms of pure carbon, is completely biocompatible. The inert chemical nature of GPC makes it useful for medical implants. Many living cells need surfaces, such as bone and collagen, to develop properly and the surface of GPC seems just as acceptable to those cells as the natural substrates.

When rapid and complete cell attachment is desirable, for sub coetaneous electrodes or temporary fluid delivery tubes, for example, GPC serves perfectly without surface treatment. However, for many applications the spatial control of cell attachment is essential [1]. The strength, durability and low density make GPC a favored material for the manufacture of artificial heart valves. Connective tissue cells naturally attach and encapsulate the implant, a desirable consequence of acceptance of the implant. However, the low cell adhesion to the glassy surfaces of the moving parts of the GPC valve has the potential [2] of creating an embolism if tissue is released into the blood stream. For the GPC heart valve, and for many applications in biology, our goal is to control the locations that cells do and do not attach.

Oxygen ion bombardment of GPC has been used [3, 4] to increase the surface roughness and enhance cell adhesion, and implanted silver ions near the surface of GPC inhibit cell attachment and adhesion [5, 6]. Both enhancement and inhibition of cell attachment to GPC are desirable modifications of GPC for particular medical applications.

We have used ion bombardment and ion implantation to provide physical and chemical signals at the GPC surfaces that either increases cell adhesion or inhibits cell attachment. We report here our progress in using patterned silver ion implantation to define areas on GPC that inhibit cell attachment, while nearby areas without silver are as acceptable to the cells as untreated GPC.

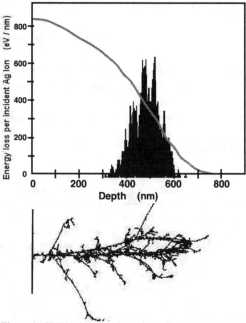

EXPERIMENTAL METHOD:

The GPC samples were prepared from phenol resin and heat treated to 700°C by methods that have been reported elsewhere [7-10]. Individual GPC samples, a few hundred microns thick and 1 x 5 cm^2 area, were exposed to a 1 MeV silver ion beam, accelerated by the AAMU Pelletron ion accelerator. The beam was scanned uniformly across the area of the GPC sample. Silver foil masks covered parts of the sample to permit patterned implantation of the fast silver ions.

Figure 1: SRIM [10] simulations of the effect of 1 MeV silver ions on GPC. The silver ions are implanted at an average depth of about 500 nm. The graph at the top shows that each incident ion deposits more energy near the surface of the GPC. The SRIM simulation below for three incident silver ions shows that recoil carbon atoms carry energy laterally hundreds of nm.

Individual samples remained exposed to the ion beam until fluences of silver ions of 10^{16} and 10^{17} cm^{-2} were accumulated. The lower fluence corresponds to 100 silver ions per nm^2. The higher fluence slightly changes the reflectivity of the GPC such that patterns are visible with the unaided eye.

Figure 1 shows the SRIM [11] simulation of the effect that individual 1 MeV silver ions have near the GPC surface exposed to the ion beam. The incident silver ions stop and remain in a diffuse layer about 100 nm thick 500 nm below the surface. Before coming to rest, each silver ion looses its energy along a meandering track. The lower half of figure 1 shows the tracks and simulated effects of three incident silver ions. The initial silver ion energy is dispersed forward and laterally by a large number of recoil carbon atoms.

We have extensive experience bombarding GPC with heavy ions [12, 13]. For these fluences, the available porosity of GPC heat treated to 700°C, already a porous material, increases a hundredfold [14]. This enhanced porosity may contribute critically to the effect of embedded silver ions on chemical signals at the surface.

Cell adhesion to GPC was tested with L929 cells of the mouse connective tissue fibroblast cell-line, from the Animal Cell Culture Collection HUKUK, Ankara, Turkey.

Standard procedures were followed [6] to determine quantitatively the degree of cell attachment, and strong cell adhesion. Optical and electronic micrographs were obtained.

RESULTS AND DISCUSSION:

Figure 2 shows the time progression of cell attachment and adhesion on untreated GPC. Following cell attachment, the spherical cells withstand gentle shear forces, whereas during the later stages the cell body becomes flat and provides strong adhesion to the substratum. The formation of spines enables the early detection of physical and chemical signals [15] that may prevent this progression to strong adhesion on inhospitable substrata. Although the cells are 10 μm diameter, the spines explore areas smaller than 100 nm [16]

Figure 3 shows a scanning electron micrograph of a border region of GPC between an implanted and an untreated area. Many such SEM images have been studied to verify the almost total inhibition of cell attachment to regions of GPC that have been implanted with a fluence of silver atoms greater than 10^{16} cm^{-2}. Evidence of strong adhesion is present when the tissue cells form the filapodial extensions displayed in the lower micrographs of untreated GPC in Figure 2. The test cells attach and strongly adhere to GPC without silver implantation.

Examination of the boundary between silver implanted and untreated areas on the same GPC surface shows that the cells adhere just as strongly within about 20 μm of that boundary as they do far from it, or as they do on a separate untreated GPC test sample.

Figure 3, and other optical and SEM images, shows an abrupt inhibition of connective tissue attachment at the edge and inside areas which are implanted with silver to fluence greater than 10^{16} cm^{-2} [5]. The evidence of strong

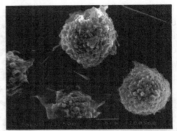
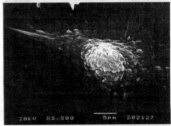

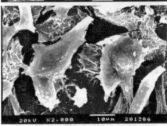

Figure 2: SEM micrographs showing the progression from tenuous cell attachment (top) to strong adhesion (bottom) on untreated GPC. The spiny filapodia are sensitive to physical and chemical signals that may inhibit this progression on inhospitable substrata.

adhesion is present among the attached cells on untreated GPC close to the implanted areas. The filapodial extensions discernible in the untreated part of the sample shown in figure 3(a) are similar to those magnified in the images of cells attached to untreated GPC shown in figure 2. We take this to mean that inhibition of cell attachment is by chemical clues at the silver implanted surfaces, and not by a toxic effect of silver in the cell solution culture, which would inhibit cell attachment everywhere.

Incident 1 MeV silver ions come to rest in a sub surface layer where they represent an atomic fraction of about 1:70 with the carbon host for an implanted fluence of 10^{16} silver ions per cm^2. We have

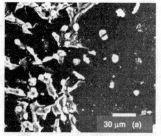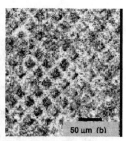

Figure 3: Inhibition of connective tissue cell attachment to GPC implanted with silver on the right half of SEM image (a). The optical image (b) shows a pattern of cell growth produced by implanting 1 MeV silver ions through a silver screen.

shown [13] that a fluence of 10^{16} ions per cm^2 significantly modifies the polymeric structure of GPC [14] in the layer of carbon between the surface and the implanted silver. Already porous (density 1.45 g/cm^3), the material becomes more permeable to fluids. The layer of silver apparently acts as a non contact electrode making the surface of GPC inhospitable to living cells.

Samples that were subjected to the conditions required to obtain SEM images were subsequently useless for additional cell growth tests, whereas samples examined optically could be cleaned and sterilized and used for additional tests. GPC samples silver implanted and tested first in July 2005 have shown persistence to inhibit cell attachment since then.

CONCLUSIONS:

Implantation of 10^{16} or more silver ions per cm^2 near the surface of carbon makes the surface of that normally biocompatible material inhospitable to L929 mouse connective tissue cells. The same cells attach and strongly adhere to untreated areas of carbon that are within a distance of 20 μm or more from an adjacent area with that fluence of implanted silver atoms. There is no evidence of a toxic effect on the cell solution, and no evidence of leaching of the implanted silver that would diminish the inhibitory effect after long term exposure to a physiological environment.

ACKNOWLEDGEMENT:

Research sponsored by the Center for Irradiation of Materials, Alabama A&M University and by National Science Foundation under Grant No. EPS-0447675.

REFERENCES:

1. S.A. Mitchell, N. Emmison and A.G. Shard, *Surface and Interface Analysis* **33** (2002) 742-747
2. N.S. Braunwald, L.I. Bonchek, *J. Thoracic & Cardiovasc. Surg.* **54-5**, 127 (1967).
3. M. G. Rodrigues, D. Ila, M. C. Resende, A. Damião and R. L. Zimmerman, "Surface Treatment of Glassy Polymeric Carbon Artifacts for Medical Applications," *Application of Accelerators in Research and Industry*, edited by Jerome L. Duggan and I. L. Morgan, (AIP 1999) 1066-1069.

4. R. Zimmerman, D. Ila, C. Muntele, M . Rodrigues, D.B. Poker and D. Hensley, "Enhanced Tissue Adhesion by Increased Porosity and Surface Roughness of Carbon Based Biomaterials", *Nuclear Instruments and Methods in Physics Research B191,* 825-829 (2002).
5. Patterning of Cell Attachment to Biocompatible Glassy Polymeric Carbon by Silver Ion Implantation, R. Zimmerman, I. Gurhan, C. Muntele, D. Ila, F. Ozdal-Kurt and B. H. Sen, *Materials Research Society Symposium II,* San Francisco, April 2006
6. R. L. Zimmerman, I. Gürhan, C. Muntele, D. Ila, M. Rodrigues, F. Özdal-Kurt and B. H. Sen, *Surface Modification of Materials by Ion Bombardment*, Izmir, Turkey, (2005).
7. G.M.Jenkins and K.Kawamura, *Polymeric Carbons-Carbons Fiber* (Cambridge University Press 1976).
8. G.M. Jenkins, C.J.Grigson, *J. Biomedical Materials Research 13*, 371 (1979).
9. H. Maleki, D. Ila, G.M. Jenkins, R.L. Zimmerman, A.L. Evelyn, *Material Research Society Symposium Proceeding 371*, 443 (1995).
10. H. Maleki, L. R. Holland, G. M. Jenkins and R. L. Zimmerman, **Carbon 35:2**, 227-234 (1997).
11. J. F. Ziegler, J. P. Biersack, *The Stopping and Range of Ions in Solids*, U. Littmark (Pergamon 1985).
12. R.L. Zimmerman, D. Ila, G.M. Jenkins, H. Maleki, D.B. Poker, *Nuclear Instruments and Methods in Physics Research B106*, 550 (1995).
13. H. Maleki, L.R. Holland, G.M. Jenkins, R.L. Zimmerman, *Journal of Material Research 11-9*, 2368 (1996).
14. R.L. Zimmerman, D. Ila, D.B. Poker, S.P. Withrow, *Application of Accelerators in Research and Industry*, Duggan & Morgan (Eds), New York, 957 (1996).
15. S. Drotleff1, U. Lungwitz1, M. Breunig1, A. Dennis1,2, T Blunk1, J. Tessmar3, A Göpferich, *European Journal of Pharmaceutics and Biopharmaceutics 85* (2004) 385 - 407
16. S. P. Massia, J.A. Hubbell, *J. Cell Biol. 114* (1991) 1089-1100.

Mater. Res. Soc. Symp. Proc. Vol. 1020 © 2007 Materials Research Society   1020-GG05-02

Nano-film and Coating for Biomedical Application Prepared by Plasma-based Technologies

Xuanyong Liu[1,2], and Paul K Chu[2]

[1]Shanghai Institute of Ceramics, Chinese Academy of Sciences, 1295 Dingxi Road, Shanghai, 200050, China, People's Republic of

[2]Department of Physics & Materials Science, City University of Hong Kong, Kowloon, Hong Kong

ABSTRACT

Nanosized materials have been widely applied in biomedical engineering due to their unique nano-effects. In this work, nano-TiO_2 coatings and ZrO_2 films were prepared using plasma technologies including plasma spraying and cathodic arc plasma deposition. The microstructure the coatings and films were assessed using TEM, SEM, and AFM. Their bioactivity and biocompatibility were evaluated using simulated body fluid soaking tests and cell culturing. Films and coatings with nanostructured surfaces can be obtained using plasma spraying and cathodic arc plasma deposition. The nanostructured surfaces can endow the films and coatings excellent bioactivity and biocompatibility. The UV-illuminated and hydrogen implanted nano-TiO_2 coatings and ZrO_2 films can induce bone-like apatite formation on their surfaces after immersion in a simulated body fluid for a certain period of time. The nano-TiO_2 coating has better cytocompatibility than the micro-TiO_2 coating, and the cytocompatibility can be improved by UV-illumination and hydrogen implantation. The bioactivity of the ZrO_2 thin film deteriorates after thermal treated. The size of the particles on the surface of the film is thought to be one of the key factors responsible for the bioactivity.

INTRODUCTION

After medical devices are implanted into the human body, interactions between the biological environment with artificial materials surfaces, onset of biological reactions, as well as particular response paths chosen by the body occur. The materials surface plays an extremely important role in the response of the biological environment to the artificial medical devices. The initial protein interactions with implant surfaces are very important, and therefore, it is clear that if the surface morphology, structure, composition, and properties are changed, cell functions are influenced. Nanoscale materials are thought to interact with some proteins more effectively than conventional materials to mediate osteoblast functions due to their similar size and altered energetics. Balasundaram et al. [1] thought nanophase materials might be an exciting successful alternative orthopedic implant materials due to their ability to mimic the dimensions of constituent components of natural bone (like proteins and hydroxyapatite).

Webster, et al. [2-4] revealed that nanophase ceramics could promote osseointegration that is critical to the clinical success of orthopedic/dental implants. Osteoblast proliferation was observed to be significantly higher on nanophase alumina, titania, and hydroxyapatite (HA) in comparison with their conventional counterparts. Furthermore, compared to conventional ceramics, synthesis of alkaline phosphatase and deposition of calcium-containing minerals were significantly enhanced on osteoblasts cultured on nanophase ceramics. Price et al. [5] showed that alumina nanometer fibers with size of 2-4 nm in diameter significantly stimulated osteoblast responses such as adhesion, alkaline phosphatase activity, and calcium deposition, when compared to alumina nanospheres with size of 23 nm in diameter and conventional spherical alumina with size of 167 nm in diameter. This nanofiber similar to the shape of hydroxyapatite crystals in bone could influence the conformation of typical adhesive proteins (that is, fibronectin and vitronectin) and osteoblast behavior [6,7].

Some novel nanostructures that have increased responses from osteoblasts leading to more efficient deposition of calcium-containing minerals include nanometals, carbon nanofibers/nanotubes, nanopolymers, and nanocomposites of ceramics and polymers. Studies were conducted on a wide range of nanomaterial chemistries and suggested that increased bone cell functions might be independent of the underlying materials chemistry and dependent only on the degree of nanostructured surface roughness [8-11].

Plasma surface modification not only chemically and/or physically alters the surface composition and microstructures via processes such as etching, chemical reactions, sterilization, ion irradiation and bombardment, but also can synthesize a different structure such as a thin film on the surface of the biomaterials via coating, grafting, and thin film deposition. The new film or surface can possess very different chemical and biological properties than the bulk materials. Among the various plasma surface modification technologies, plasma spraying and plasma immersion ion implantation and deposition (PIII&D) have been widely used to modify the surface structure, composition and properties of biomaterials [12].

The nanoscale surface is thought to have more promising biological properties as discussed above. Therefore, in this work, nano-TiO_2 coatings and ZrO_2 films were prepared using plasma spraying and cathodic arc plasma deposition respectively, and their structures and properties were evaluated using various methods.

EXPERIMENT

Commercially available nanometer-sized TiO_2 powders with size of about 30 nm (P25, Degussa, Germany) and submicrometer-sized TiO_2 powders with size of 0.3 μm (Wuhan Institute of Materials Protection, China) were made into sphere-like particles of 45~90 μm using a spray drying process. The nanosized TiO_2 powders are composed of about 80% anatase and 20% rutile phases. The particles were deposited onto Ti-6Al-4V substrates by atmospheric plasma spraying (APS) under modified spraying parameters. The process uses an electrical arc to melt and spray materials onto a surface as illustrated shown in Fig. 1. In our experiments, the coatings produced using nano- and submicro- powders were denoted by the nano-TiO_2 and micro-TiO_2 coatings, respectively. The as-sprayed coatings were post-treated by UV-illumination in air for 24 hours and hydrogen implantation using a plasma immersion ion implanter to enhance their bioactivity.

The ZrO_2 thin films with nano-sized surface were fabricated on n-type, 100 mm Si (100) wafers using a filtered cathodic arc system [14]. The experimental apparatus used in this study included a magnetic duct and cathodic arc plasma source, as shown in Figure 2. The zirconium discharges were controlled by the main arc current between the cathode and anode. Oxygen gas was bled into the arc region and the mixed zirconium and oxygen plasma was guided into the vacuum chamber by an electromagnetic field applied to the curved duct. The duct was biased to -20 V to exert a lateral electric field while the external solenoid coils wrapped around the duct produced the axial magnetic field with a magnitude of 100 G. Before deposition, the samples that were positioned about 15 cm away from the exit of the plasma stream were cleaned by argon plasma for 2 min using a sample bias of -500 V. The base pressure in the vacuum chamber was about 1×10^{-5} Torr and RF power of 100 W was applied for a deposition time of 120 min. The as-deposited ZrO_2 thin films were thermal treated at $1000^\circ C$ for 2 hours.

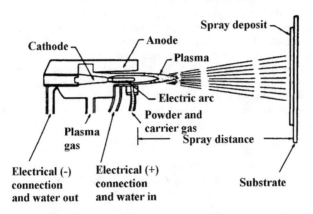

Figure 1. Schematic diagram of a plasma spray torch [13] used to prepare TiO_2 samples.

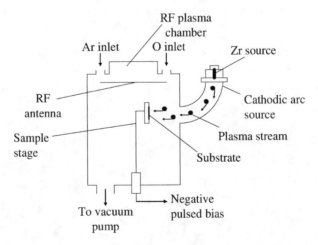

Figure 2. Schematic diagram of the magnetic duct and cathodic arc plasma source used to prepare ZrO2 samples

The surfaces of all samples were observed using cold field-emission scanning electron microscopy (SEM) using a JEOL JSM-6700F. Contact mode atomic force microscopy (AFM) was conducted on a SPI3800N & SPA300HV System (Seiko Instruments Inc., Japan) to evaluate the surface morphology of the ZrO_2 thin film over a scanned area of 2 μm × 2μ m.

After ultrasonically washed in acetone and rinsed in deionized water, all samples were soaked in simulated body fluids (SBF) [15] for 28 days at 36.5°C without stirring to investigate their bioactivity. The human osteoblasts were seeded on the titania coatings at a density of 4000 cells/cm^2. After culturing for 1, 4, 7 days in a 5% CO_2 incubator at 37°C, cell proliferation on the titania coatings were evaluated using MTT assay as reported in the literature [16]. The MTT assay is a simple and useful tool for evaluating cell vitality and proliferation. The key component is 3-(4,5-dimethylthiazol-2-yl)-2,5-diphenyltetrazolium bromide. Mitochondrial dehydrogenases of living cells reduce the tetrazolium ring, yielding a blue formazan product which can be measured spectrophotometrically. The amount of formazan produced is proportional to the number of viable cells present. MTT (5 mg ml^{-1} in DMEM without phenol red) was added to the wells in an amount equivalent to 10% of the culture medium. After an incubation of 4 h at 37°C, the liquid was aspirated and the insoluble formazan produced was dissolved in isopropanol. The optical densities were measured at 570 nm.

DISCUSSION

The high magnification surface views of the as-sprayed TiO_2 coatings displayed in Fig. 3 indicate that the surface of the nano-TiO_2 coating comprises particles about 50 nm in size (Fig. 3a), whereas the surface of the micro-TiO_2 coating is made of particles with size of about 100 nm (Fig. 3b). Our previous work revealed that the thickness of the outer layer with nano-sized particles in nano-TiO_2 coating is about 500 nm [17]. In the interior of the coating, most of the grains exhibit a columnar morphology with a diameter of about 100~200 nm. The difference in the crystal growth between the surface and interior of the coating depends mostly on the thermal history. During plasma spraying, the bulk of the coating tends to possess larger columnar grains due to the continuous heat provided by the plasma and subsequent melt, whereas the surface grains are subjected to less heating.

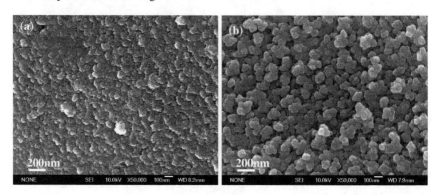

Figure 3. Surface SEM views of as-sprayed TiO_2 surfaces: (a) nano-TiO_2 coating and (b) micro-TiO_2 coating [17].

After immersion in SBF for four weeks, the surface of the hydrogen implanted and UV-illuminated nano-TiO$_2$ coating was completely covered by bone-like apatite [17]. In contrast, in our control experiments involving the as-sprayed nano-TiO$_2$ and micro-TiO$_2$ coatings as well as the hydrogen implanted or UV-illuminated micro-TiO$_2$ coating and hydrogen implanted or UV-illuminated polished nano-TiO$_2$ coating with the nanostructured surface removed, no apatite could be observed on the surfaces after they were soaked in SBF for four weeks. Our results indicate that only the hydrogen implanted and UV-illuminated nano-TiO$_2$ coatings with nanostructured surfaces possess bone-like apatite formability. It can thus be inferred that the bioactivity of the plasma-sprayed TiO$_2$ coating depends on nanostructured surface composed of enough small particles and hydrogen incorporation or UV-illumination.

The MTT assay results of the osteoblasts cultured on the different TiO$_2$ coatings for various time are shown in Fig. 4. It can be observed that the osteoblasts seeded on the surface of the nano-TiO$_2$ coating proliferate faster than that on the micro-TiO$_2$ coating, and more cells can be detected on the UV-illuminated and hydrogen implanted nano-TiO$_2$ coatings than on the as-sprayed nano-TiO$_2$ coating. The results indicate that the nano-TiO$_2$ coating has better cytocompatibility than the micro-TiO$_2$ coating, and the cytocompatibility can be improved by UV-illumination and hydrogen implantation.

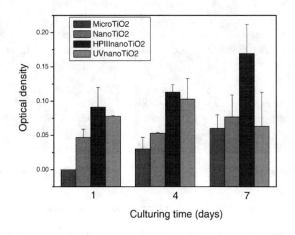

Figure 4. MTT assay results of the cells seeded on various TiO$_2$ coating for different period

The surface views of the as-deposited and thermally treated ZrO$_2$ thin films observed by SEM and AFM are displayed in Figs. 5 and 6. The surface of the as-deposited ZrO$_2$ thin film is very smooth and the surface features cannot be distinguished very well by SEM, as shown in Fig. 5a. However, some very small particles can be observed on the surface of the as-deposited ZrO$_2$ thin film in the AFM picture (Fig. 6a). After the ZrO$_2$ thin film was annealed at 1000°C for 2 hour, the particles on its surface grew to about 30~50 nm, as shown in Figs. 5b and 6b.

Figure 5. SEM views of the ZrO_2 thin film: (a) as-depositedand and (b) treated at $1000^{\circ}C$ for 2h.

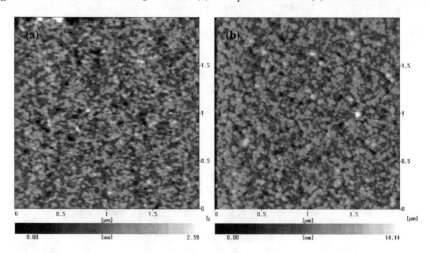

Figure 6. AFM views of the ZrO_2 thin film: (a) as-depositedand and (b) treated at $1000^{\circ}C$ for 2h.

The as-deposited and thermal-treated treated ZrO_2 thin films were immersed in SBF to evaluate their bioactivity. The surface views of the as-deposited and thermal-treated treated ZrO_2 thin films soaked in SBF for 28 days are depicted in Fig. 7. After immersion in SBF for 28 days, the surface of the as-deposited ZrO_2 thin film was completely covered by the apatite layer, while relatively few apatites appeared on the surface of the ZrO_2 thin film annealed at $1000^{\circ}C$ for 2h. The results indicate that the bioactivity of the ZrO_2 thin film degrades after thermal treatment. The size of the particles on the surface of the film is thought to be one of the key factors to its bioactivity. It is also believed to be one of the reasons why the ZrO_2 thin film is bioactive while other ZrO_2 materials such as ZrO_2 ceramic [18] and coatings [19] are bioinert.

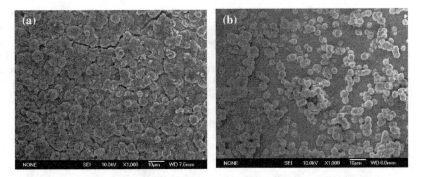

Figure 7. Surface views of the different ZrO_2 thin film soaked in SBF for 28 days: (a) as-deposited and (b) thermal-treated at $1000^{\circ}C$ for 2h.

The results obtained from the nano-TiO_2 coating and ZrO_2 thin film may indicate that the nano-sized surface bodes well for bioactivity and biocompatibility compared to a conventional surface. The deposition of calcium ions is the first and most crucial step of carbonate-containing hydroxyapatite nucleation from an ionic solution. This process is believed to initiate the growth of bone-like apatite on the surface of biocompatible implants [20]. The formation of a negatively-charged surface gives rise to apatite precipitation because positive calcium ions are attracted from the solution [21]. The charge densities of the particles are determined by its size. Vayssieres et al. [22] have suggested that finer nanocrystalline particles have higher surface charge densities than larger ones. It can also be demonstrated by thermodynamic analysis that the surface or interfacial tension diminishes with decreasing particle size as a result of the increase in the potential energy of the bulk atoms in the particles [23]. Smaller particles with increased molar free energy are more likely to adsorb molecules or ions onto their surfaces in order to decrease the total free energy and to become more stable. Therefore, the nano-sized particles in the outermost layer of the TiO_2 coating and ZrO_2 thin film may be the key factor inducing the precipitation of bone-like apatite on the surface during immersion in SBF.

Some researchers reported that the biocompatibility of the materials was improved by the nano-sized surface. G. Balasundaram et al. suggested that decreasing the particulate size into the nanometer regime and reducing crystallinity of calcium phosphate based materials may promote osteoblast adhesion. The proteins adsorbing onto nanophase ceramics can enhance osteoblast adhesion and cause immobilization of arginine–glycine–aspartic acid (RGD) on conventional ceramics. [24]. Webster et al. thought that one of the main reasons why nanophase materials attracted selected proteins to their surface was due to their altered surface energetics compared to conventional materials (>100 nm). Several other studies have highlighted altered protein bioactivity when adsorbed to nanophase compared with conventional materials [25,26].

CONCLUSIONS

TiO_2 coatings and ZrO_2 films with nano-sized surfaces were prepared using plasma spraying and cathodic arc plasma deposition, respectively. The nanostructured surfaces give rise to excellent bioactivity and biocompatibility. The bioactivity very much depends on a nanostructured surface composed of enough small particles. The UV-illuminated and hydrogen implanted nano-TiO$_2$ coatings and ZrO$_2$ films can induce bone-like apatite formation on their surfaces after immersion in a simulated body fluid for a certain period of time. The nano-TiO$_2$ coating has better cytocompatibility than the micro-TiO$_2$ coating, and the cytocompatibility can be improved by UV-illumination and hydrogen implantation. The bioactivity of the ZrO$_2$ thin film deteriorates after thermal treated. The size of the particles on the surface of the film is thought to be one of the key factors responsible for the bioactivity.

ACKNOWLEDGMENTS

This work was jointly supported by National Basic Research Fund under grant 2005CB623901, Shanghai Science and Technology R&D Fund under grant 0552nm014, Foundation for the Author of National Excellent Doctoral Dissertation of PR China (FANEDD), and Hong Kong Research Grants Council (RGC) Competitive Earmarked Research Grant (CERG) No. 112306.

REFERENCES

1. G. Balasundaram and T. J. Webster, *J. Mater. Chem.* 16, 3737-3745 (2006).
2. T. J. Webster, R. W. Siegel and R. Bizios, *Biomaterials* 20, 1221-1227 (1999).
3. T. J. Webster, C. Ergun, R. H. Doremus, R. W. Siegel and R. Bizios, *Biomaterials* 21, 1803-1810 (2000).
4. T. J. Webster, C. Ergun, R. H. Doremus, R. W. Siegel and R. Bizios, *Biomaterials* 22, 1327-1333 (2001).
5. R. L. Price, L. G. Gutwein, L. Kaledin, F. Tepper and T. J. Webster, *J. Biomed. Mater. Res.* 67A, 1284-1293 (2003).
6. M. M. Stevens and J. H. George, *Science*, 310, 1135-1138 (2005).
7. C. J. Wilson, R. E. Clegg, D. I. Leavesley and M. J. Pearcy, *Tissue Eng.* 11, 1-18 (2005).
8. T. J. Webster and J. U. Ejiofor, *Biomaterials* 25, 4731-4739 (2004).
9. R. L. Price, M. C. Waid, K. M. Haberstroh and T. J. Webster, *Biomaterials* 24, 1877-1887 (2003).
10. T. J. Webster and T. A. Smith, *J. Biomed. Mater. Res.* 74A, 677-686 (2005).
11. I. Manjubala, S. Scheler, J. Bossert and K. D. Jandt, *Acta Biomater.* 2, 75-84 (2006).
12. P. K. Chu, J. Y. Chen, L. P. Wang, and N. Huang, *Mater. Sci. Eng. R* 36, 143-206 (2002).
13. M.I. Bouos, P. Fauchais and A. Vardelle, in *Plasma Spraying: Theory and Application*, edited by R. Suryanarayanan, (World Scientific, Singapore, 1993) pp. 3.
14. T. Zhang, P. K. Chu and I. G. Brown, *Appl. Phys. Lett.* 80, 3700-3702 (2002).
15. T. Kokubo, H. Kushitani, S. Sakka, T. Kitsugi and T. Yamamuro, *J. Biomed. Mater. Res.* 24, 721-734 (1990).
16. A. Oliva, A. Salerno, B. Locardi, V. Riccio, F. D. Ragione, P. Iardino, and V. Zappia, *Biomaterials* 19, 1019-1025 (1998).
17. X. Y. Liu, X. B. Zhao, R. K. Y. Fu, J. P. Y. Ho, C. X. Ding, and P. K. Chu, *Biomaterials* 26, 6143-6150 (2005).

18. M. Ferraris, E. VerneH, P. Appendino, C. Moisescu, A. Krajewski, A. Ravaglioli and A. Piancastelli, *Biomaterials* 21, 765-773 (2000).
19. X. Y. Liu, C. X. Ding, *Surf. Coat. Tech.* 172, 270-278 (2003).
20. M. Svetina, L. C. Ciacchi, O. Sbaizero, S. Meriani and A. de Vita, *Acta Mater.* 49, 2169-2177 (2001).
21. P. Li, C. Ohtsuki, T. Kokubo, K. Nakanishi, N. Soga and K. de Groot, *J. Biomed. Mater. Res.* 28, 7-15 (1994).
22. L. Vayssières, C. Chanéac, E. Trone and J. P. Joliver, *J. Colloid. Interface Sci.* 205, 205-212 (1998).
23. H. Zhang, R. L. Penn, R. J. Hamers and J. F. Banfield, *J. Phys. Chem. B* 103, 4656-4662 (1999).
24. G. Balasundaram, M. Sato and T. J. Webster, *Biomaterials* 14, 2798-2805 (2006).
25. T. J. Webster, in *Advances in Chemical Engineering*, edited by J. Ying, (Academic Press Inc., CA, 2003) pp. 125–166.
26. T. J. Webster, C. Ergun, R. H. Doremus, R. W. Siegel and R. Bizios, *J. Biomed. Mater. Res.* 51, 475-483 (2000).

Mater. Res. Soc. Symp. Proc. Vol. 1020 © 2007 Materials Research Society   1020-GG05-03

Modification of Surface Morphology of UHMWPE for Biomedical Implants

Ahmet Oztarhan[1], Emel Sokullu Urkac[1], Nusret Kaya[1], Mesut Yenigul[2], Funda Tihminlioglu[3], Ayhan Ezdesir[4], Robert Zimmerman[5], Satilmis Budak[5], C Muntele[5], Bopha Chhay[5], Daryush Ila[5], Efim Oks[6], Alexey Nikolaev[6], Zekai Tek[7], and Rengin Eltem[1]

[1]Bioengineering, Ege University, Bornova, Izmir, 35100, Turkey
[2]Chemical Engineering, Ege University, Bornova, Izmir, 35100, Turkey
[3]Chemical Engineering, Izmir Institute of Technology, Izmir, 35000, Turkey
[4]Petkim, Izmir, 35000, Turkey
[5]Alabama A&M University, Huntsville, AL, 35762
[6]High Current Electronics Institute, Tomsk, 634055, Russian Federation
[7]Celal Bayar University, Manisa, 45000, Turkey

Abstract

Ultra High Molecular Weight Polyethylene (UHMWPE) samples were implanted with metal and metal-gas hybrid ions (Ag, Ag+N, C+H, C+H+Ar, Ti+O) by using improved MEVVA Ion implantation technique [1,2]. **An extraction voltage of 30 kV and influence of 1017 ions/cm2 were attempted in this experiment.** to change their surface morphologies in order to understand the effect of ion implantation on the surface properties of UHMWPEs. Characterizations of the implanted samples with RBS, ATR - FTIR, spectra were compared with the un-implanted ones. Implanted and unimplanted samples were also thermally characterized by TGA and DSC. It was generally observed that C–H bond concentration seemed to be decreasing with ion implantation and the results indicated that the chain structure of UHMWPE were changed and crosslink density and polymer crystallinity were increased compared to unimplanted ones resulting in increased hardness. It was also observed that nano size cracks (approx.10nm) were significantly disappeared after Ag implantation, which also has an improved antibacterial effect. Contact angle measurements showed that wettability of samples increased with ion implantation. Results showed that metal and metal+gas hybrid ion implantation could be an effective way to improve the surface properties of UHMWPE to be used in hip and knee prosthesis.

1. Introduction

Ultra High Molecular Weight Polyethylene (UHMWPE) has been commonly used for acetabular cup of total hip joint replacement as it has densely packed linear polyethylene chains, which gives improved mechanical properties. However the wear of UHMWPE against the articulating metal part and wear debris generated at the surface is recognized as the major cause of loosening and failure of the total joint replacement. Recent studies show that increasing the cross-linking in the polyethylene significantly reduces wear leading to more durable acetabular components and increasing the lifetime of an implant. Increasing the cross-link density of UHMWPE in which UHMWPE is irradiated in air at an elevated temperature with a high-dose-rate electron beam or a common approach is to cross-link the polymer by gamma irradiation, which improves markedly the wear resistance of the polymer but with changed bulk properties [3]. In this work, we tried to change the surface morphology of UHMWPE by low energy metal and metal+gas hybrid ion implantation by MEVVA , with an expectation of improving surface mechanical properties, and promoting antibacterial effect without changing the bulk properties of UHMWPE .

2. Experimental
Samples with medical grade GUR 1020 - Type 1 - Ultra High Molecular Weight Polyethylene (UHMWPE - , Hipokrat Co.) with a density of 945 kg/m^3 were used . Disk samples with a diameter of 30mm and thickness of 4mm were polished down to about surface roughness of 95 (nm) Ra. Samples were implanted with Ag, Ag+N, C+H, C+H+Ar, Ti+O ions by using MEVVA ion implanter with a fluence of 10^{17} ion/cm^2 and extraction voltage of 30 kV. ATR- FTIR chemical characterization analysis was used to see if any new chemical bonds formed 2 microns deep at the surface. Thermo Nicolet Nexus 670 model FTIR with Smart DuraSampllR 3 Bounce diamond HATR (3 reflection diamond ATR) and OMNIC software were used. DSC (Differential Scanning Calorimeter) SHIMADZU DT-50 and the TGA (Thermo Gravimetric Analyzer) SHIMADZU DT-51 were used for thermal analysis of implanted and unimplanted samples. The analyses were performed in a dry nitrogen atmosphere. Aluminum cells were used for analysis of the polymers. The temperatures of the sample cells were increased by 10 $^0C/min$. The total heat of melting ΔH (the area under the endotherm-DSC) was determined and, knowing the total heat of fusion of 100% crystalline UHMWPE ($\Delta H_f = 293.6$ J/g), the percentage crystallinity was calculated as 100($\Delta H/\Delta H_f$) [4]. Scanning electron microscopy (Philips XL-305 FEG – SEM) was used to examine the surfaces of unimplanted and Ag and Ag+N implanted samples. Atomic force microscopy (AFM) was used to investigate the surface morphology of unimplanted and implanted UHMWPE. A Digital Instrument- MultiModeSPM apparatus was used to determine the surface roughness. The durometer hardness test was used to measure the relative hardness of the samples. The test method is based on the penetration of a specified indenter forced into the material under specified conditions. A Shore hardness tester ZWICk/Roell (HPE) under 50N force was used. The contact angle of wateron unimplanted and implanted UHMWPE surfaces was measured with a Krüss-G10 goniometer. The antibacterial activities of unimplanted and Ag implanted UHMWPE samples of 5mm diameter and 3mm thickness were examined against "Staphylococcus Aureus" by "Agar Disk Diffusion Method". After 24 hours bacteria reduction was100%.

3. Results and Discussions
RBS graphs, are used as an analytical tool to measure some of the implanted elements (C, Ag, Ti, N, O) concentration of the sample surfaces. For example, Figure 1,2 show Ag, Ti, C , N and O ions, which can be detected up to 42 +15 nm underneath the surface. All spectra show an O peak, which is believed to be caused by oxidation and some small peaks by the residual air in the Vacuum chamber. Oxygen uptake takes place once the ion bombarded polymer sample was exposed to the air [5].

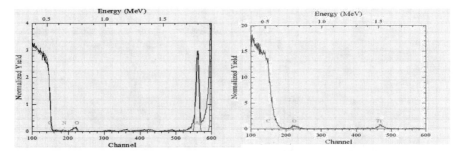

Figure.1 RBS Spectrum of Ag + N
Implanted UHMWPE

Figure.2 RBS Spectrum of Ti + O
Implanted UHMWPE

The characteristic absorption bands for the CH_2 bonds appear in the 2900–2840, 1460–1370 and 740–720 cm^{-1} regions [6]. The transmission ATR analysis of the unimplanted and implanted samples confirms the C–H bond breaking since the C–H stretching (at 2847 cm^{-1}) bond peaks of the pure UHMWPE sample disappear after ion implantation (figures. 3,4,5).

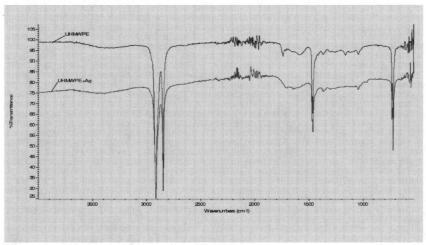

Figure.3 FTIR spectra of Ag Implanted UHMWPE

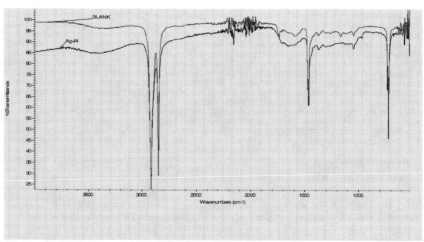

Figure.4 FTIR spectra of Ag+N Implanted UHMWPE

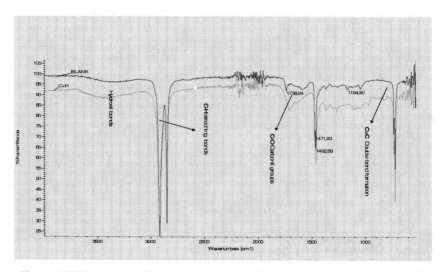

Figure.5 FTIR spectra of C+H Implanted UHMWPE

Increase in the absorption bands in the 800 and 1738.4 cm^{-1} regions , due to the implantation, has been attributed to the creation of unsaturated C=C bonds and carbonyl group formations respectively [7] . One of the main products of the ion bombardment is the chemical change resulting the C=C formation, and its stretching vibration is clearly observed in the spectra by an absorbance peak at 800 cm^{-1}. The presence of this peak suggests that after ion implantation, the polymer surface becomes poor of hydrogen and rich of unsaturation which is susceptible to crosslinking and oxidation. Oxidation causes carbonyl groups formation which is observed at around 1740 cm^{-1}. DSC (Differential Scanning Calorimeter) was used to measure the energy change of the polymer by increasing the temperature in inert N_2 gas atmosphere. DSC determines thermal properties of the polymer such as T_m, ΔH_{fus}. ΔH_{fus} is defined as the fusion enthalpy which is calculated by taking an integration of the area of the melting peak. The crystallinity of UHMWPE (unimplanted and implanted) percentage is shown in table.1, tohether with roughness, hardness, contact angle, ΔH_{fus} and T_m. The detailed calculation of the crystallinity percentage of the polymer could be found in the literature [4]. DSC Analysis of un implanted UHMWPE, Ag and Ag+N, and C+H+Ar and C+H ion implanted UHMWPE samples are shown in figures 6 and 7.

Table gives the contact angle and roughness results together with the other measured properties. It seems that the wettability increased with ion implantation and that hidroxil bonds and carbonil bonds have dominant effect on wettability

	Roughness (Ra-nm)	Hardness (Shore-D)	Contact Angle($^\circ$)	ΔH_f J/g)	Tm (C^0)	Crystallinity (DSC - %)
Untreated	124	58	56	115.07	141	39,5
Ag	30	66.4	45	116.52	137	40
Ag+N	74	63.4	32	119.67	140	41
C+H	69.5	64,2	37.5	115.82	135.2	39,9
C+H+Ar	101	66,1	47.5	117.57	131.8	40,5

Table 1 . Over all results

Table.1 Over all results

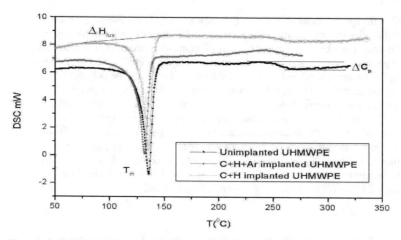

Figure .6. DSC Analysis of unimplanted UHMWPE, C+H+Ar and C+H ion implanted UHMWPE samples.

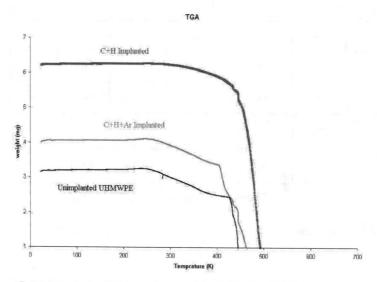

Figure .7. DSC Analysis of Untreated, Ag and Ag+N Implanted UHMWPE Samples

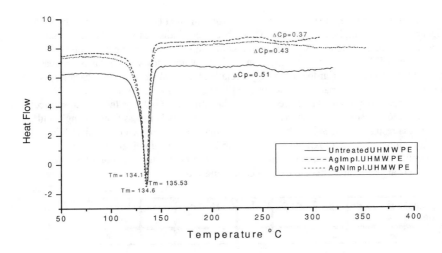

Figure . 8. TGA curves of the samples of UHMWPE before and after the ion implantation with C+H, C+H+Ar ions.

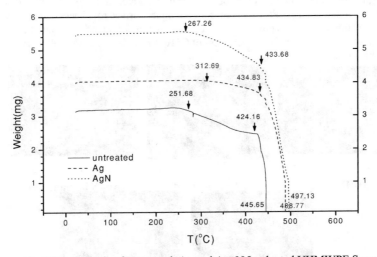

Figure . 9. TGA Analysis of Untreated, Ag and Ag+N Implanted UHMWPE Samples.

In TGA results, onset of degradation starts for untreated UHMWPE is 251.68 ^{0}C. This point shifted after irradiation to higher values for implanted samples (Figures 8, 9). With the same trend total degradation point 424 ^{0}C for untreated UHMWPE sample shifted to higher values for implanted samples. These results can indicate that the thermal stability increased at the surface [8] . Shore-D hardness measurements of unimplanted and implanted UHMWPE samples are given in Table.1 , which shows that hardness increases with ion implantation . Hardness is purely a relative term and an increase in hardness is due to the crosslink formation on the surface [9,10]. It was also observed that C–H bond concentration seemed to be decreasing. This might also indicate that C rich surfaces and graphite like structures occurred, resulting in increased hardness.Contact angle and roughness results are given in Table.2 . It seems that the wettability increased with ion implantation and that hidroxil bonds and carbonil bonds have dominant effect on wettability. Rather low contact values for Ag+N could be due to formation of nitride which has an additional effect to oxygen containing groups.

Contact Angle	UHMWPE	Ag	Ag + N	C+H	C+H+Ar	Ti+O	Ti+Al+N
Water	56.2°	44.8°	32°	37.5	47.5	30°	48°

Table.2 Contact angle measurements with water.

Figure. 10. shows the surface topography of unimplanted and implanted UHMWPE samples obntained by AFM. As seen from the figure surface roughness decreases with ion implantation. This is probably due to rapid melting of the polymer surfaces as a result of ion irradiation. High roughness of C+H+Ar implanted UHMWPE surface could be due to Ar ion sputtering.

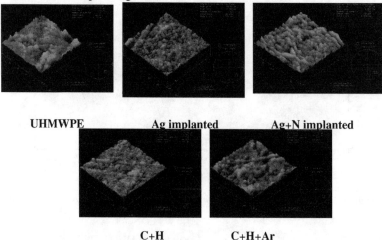

UHMWPE   Ag implanted   Ag+N implanted

C+H   C+H+Ar

Figure. 10. Surface topography of unimplanted and implanted UHMWPE samples obntained by AFM.

SEM micrographs of unimplanted, Ag and Ag+N implanted UHMWPE surfaces were obtained randomly with several measurements are shown in figure 11 . Results represent the surface morphology of UHMWPE before and after implantation. Cracks on the surface are significantly disappeared after Ag and Ag+N implantation. Especially, cracks were totally disappeared with Ag implantation.

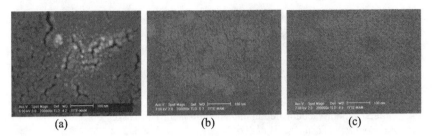

(a)   (b)   (c)

Figure 11. SEM micrographs of (a) unimplanted, (b) Ag+N implanted and (c) Ag implanted UHMWPE samples.

The results proved the Antibacterial effect of Ag implanted in UHMWPE, however, the measurements should be repeated after the wear tests.

Conclusion

The results indicated that the chain structure of UHMWPE were changed and crosslink density increased. There was slight increase in percentage crystallinity but decrease in melting point T_m with irradiation, which may be due to the chain sission of UHMWPE since T_m is closely related to number average molecular weight and also C=C double bond formation as seen in FTIR spectra.
It is believed that slight increase in the crystallinity of the polymer may improve the surface quality of the polymers such as hardness, wear resistance and wettability.
It was also observed that C–H bond concentration seemed to be decreasing. This might indicate that C rich surfaces and graphite like structures occurred, resulting in increased hardness. It was observed that nano size cracks (approx.10nm) significantly disappeared after Ag implantation, however less significant with Ag+N implantation . This is probably due to rapid melting of the polymer surfaces as a result of ion irradiation, which was also thought to be responsible for decreasing in roughness.
Contact angle measurements showed that wettability of samples increased with ion implantation. It is thought that increased in oxygen concentration and formation of carbonyl groups were responsible for this. Rather low contact values for Ag+N could be due to formation of nitride which has an additional effect to oxygen containing groups.
Results indicate that metal and metal+gas hybrid ion implantation could be an effective way to improve the surface of UHMWPE together with antibacterial effect of Ag implantation.

Acknowledgements

This work was supported by the Center for Irradiation of Materials, Alabama A&M University.

List of References

1. E. M. Oks, G. Yu.Yushkov, P.J.Evans, A.Oztarhan, I.G.Brown, M.R.Dickinson, F.Liu, R.A. MacGill, O.R. Monteiro and Z.Wang, "Hybrid gas-metal co-implantation with a modified vacuum arc ion source", *Nucl. Instr. and Meth. B, Vols 127-128, pp. 782-786, 1997.*
2. A. Oztarhan., I. Brown, C. Bakkaloglu, G.Watt, P. Evans, E.Oks, A. Nikolaev and Z.Tek, , "Metal vapour vacuum arc ion implantation facility in Turkey", *Surface and Coatings Technology, Vol 196, Issues 1-3, pp. 327-332,* 2005
3. Gamma-irradiated cross-linked polyethylene in total hip replacements - analysis of retrieved sockets after long-term implantation, Hironobu Oonishi [1*], Yoshinori Kadoya [2], Shingo Masuda [3] , Applied Biomaterials Volume 58, Issue 2 , Pages 167 – 171, Published Online: 17 Jan 2001Copyright © 2001 John Wiley & Sons, Inc.)
4. S.M. Kurtz, The UHMWPE Handbook: Principles and Clinical applications in total joint replacement , Elsevier , Academic Press, 2004.5. Turos A., Abdul-Kader AM, Grambole D, Jagielski J, Piątkowska A, Madi NK and Al-Maadeed M, 2006, "The effects of ion bombardment of ultra-high molecular weight polyethylene", Nucl. Instr. and Meth. B, Vol. 249, Issues 1-2 , pp. 660-664
6. A.L. Evelyn, D. Ila , R.L Zimmerman, K. Bhat, D.B. Poker and Hensley, 1998, "Effects of MeV ions on PE and PVDC" , Nucl. Instr. and Meth. B, Vol 141, Issues 1-4 , pp. 164-168
7. P. Bracco, Brach del Prever EM, M. Cannas, M.D. Luda and L. Costa, 2006, "Oxidation behaviour in prosthetic UHMWPE components sterilised with high energy radiation in a low-oxygen environment", Polymer Degradation and Stability, Vol 91, Issue 9 , pp. 2030-20388. Marco Zanetti, Pierangiola Bracco and Luigi Costa Polymer Degradation and Stability Volume 85, Issue 1 , July 2004, Pages 657-665
9. S.M. Kurtz, O.K. Muratoglu, M. Evans and A.A. Edidin, *Biomaterials* 20 (1999), p. 1659.
10. A. Valenza, A. M. Visco, L. Torrisi and N. Campo, Polymer, Volume 45, Issue 5 , March 2004, Pages 1707-1715.

Mater. Res. Soc. Symp. Proc. Vol. 1020 © 2007 Materials Research Society   1020-GG06-01

Ion Beam-induced Quantum Dot Synthesis in Glass

Harry Bernas[1], and Roch Espiau de Lamaëstre[1,2]
[1]Bat. 108, CSNSM-CNRS, University Paris-Sud 11, Orsay, 91405, France
[2]Corning SA, Fontainebleau Research Center, Avon, 77210, France

ABSTRACT

Irradiation-induced processes are often considered only in their nonequilibrium aspects. The purpose of this brief review is to show that chemistry, and particularly redox properties, play a major role in the thermal evolution of such systems and generally cannot, therefore, be neglected. This is exemplified by the synthesis of Ag nanoclusters in glasses and silica, under both low (gamma-ray) and high (MeV ion) deposited energy density irradiation conditions. The nanocluster formation mechanism is shown to be similar to the latent image formation process in photography. The corresponding information was used to control nucleation and growth of PbS clusters in glasses, leading to promising optical properties. In the course of these studies, we also showed that lognormal size distributions characterize the absence of information on the nanocluster formation process.

INTRODUCTION

To what extent do nonequilibrium processes circumvent familiar thermodynamical laws, and allow us to design novel ways of synthesizing nanocrystals in a given host ? This question is necessarily present when attempts are made to expand the range of synthesis techniques beyond the impressive capacities of, say, colloidal chemistry. The latter may provide a large variety of pure or composite (alloy, core-shell) nanocrystals, with very narrow size distributions (typically a few percent), but their major limitation, in some important cases, is the difficulty of inserting the nanocrystals into an appropriate host without modifying (or destroying) them. Hence the efforts to promote synthesis in the desired host, and the subsequent problems of dealing with temperature-dependent solubility, diffusion, and preferential interactions in the case of multicomponent systems.

Ion implantation (or irradiation), combined with post-annealing, has been used for many years [1] as a means of both forcing solubility and synthesizing nanocrystals in the glasses and other insulators on which we focus here. As a rule, results thus obtained have been discussed in terms of specific ion-beam interactions such as impurity-defect interactions, ion beam or recoil mixing, etc. Although chemical effects have been evidenced previously, they had not previously been related systematically to charge carrier properties. In this Symposium talk, we review our efforts in this direction, as well as a study of the meaning of lognormal nanocluster size distributions. This work has been described in a series of submitted or published papers (see references) whose main results are very briefly summarized here.

BEAM-INDUCED NANOCRYSTAL NUCLEATION AND GROWTH IN INSULATORS: ROLE OF CHARGE CARRIERS

Forcing solubility of, say, a metal element such as Ag in an insulator (e.g., a glass) or a semiconductor by ion implantation results in a compositional change, of course, and annealing leads to precipitation if supersaturation occurs. However, a "binary mixture" picture is generally not relevant to analyze the result. Besides possible irradiation-assisted (or -induced) diffusion or precipitation, an ubiquitous effect associated to any ionizing particle irradiation - including photons - is charge carrier (electron and hole) injection. Subsequent evolutions are similar to latent image formation in photography: irradiation-induced electrons and holes are trapped within the insulating matrix, notably on dissolved metal ions, which are thus neutralized (by electrons) or further oxidized (by trapped holes), and on defects which may "store" the charges until annealing releases them. Annealing also promotes metal atom (or ion) diffusion, which in turn leads to formation of cluster nuclei. The cluster density depends not only on the irradiation fluence, but also – primarily – on the density of deposited energy *and the redox properties of the insulator*. Although the influence of chemistry has been noted [2], the latter feature has often been neglected so far in spite of the fact that both defects and chemical species may exist in more than one charge state, and that reaching charge equilibrium is often a comparatively long-term process. Hence, a complete analysis of precipitation (or nanoclustering) in insulators or semiconductors should take into account (i) the density of irradiation-induced electrons and holes, (ii) the density and nature (electron- or hole-trapping) of defects, (iii) the redox potential of the various ionized or neutral states of the precipitating chemical species versus that of the host, and (iv) the transport (diffusion) properties, both of the charges and of the different metal charge states. To pursue the analogy with photography, note that photons or other ionizing particles obviously create an equal number of electrons and holes: in order to enhance Ag precipitation in the AgBr crystals of a photographic emulsion, a chemical developer is used. It is essentially an electron donor, biasing the electron- over the hole-population, thus favoring the $Ag^+ + e^- \rightarrow Ag^0$ reaction over the reverse $Ag^0 + h^+ \rightarrow Ag^+$ reaction which would allow AgBr to re-form.

How does this apply to ion beam-induced nanocluster synthesis ? We performed [3] an experimental study, via electron spin resonance (ESR) and optical absorption (OA), of Ag-containing glass or silica irradiated by gamma photons or MeV heavy ions. The values of Deposited Energy Density (DED) corresponding to these irradiations differ by about four orders of magnitude. Of course, whereas gamma irradiation affects the totality of our 1 mm-thick samples, MeV ions only affect about 3 μm. Thus, in the first case a variety of Ag species (neutral atoms, single ions or different charged oligomers, small clusters) and defects are detected; in the latter instance, irradiation defects are detected by ESR and Ag nanoclusters are seen in OA via the surface plasmon resonance excitation. The main results are the following: (1) the vast majority of metastable defects seen after room temperature gamma irradiation (weak DED) of the pure glass host (no Ag), according to ESR, are Non-Bonding Oxygen Hole Centers (NBOHC), i.e., hole traps. This contrasts with the effect of the large-DED fast heavy ion irradiation, which produces more metastable electron traps (E' defects) than hole traps. Thus, an increase in DED leads to an enhanced E'/NBOHC ratio: ion irradiation produces a highly reducing medium, favoring Ag nanocluster formation; (2) ESR experiments on Ag-containing glass at 80 K confirm this result - specifically, they evidence two series of reactions as the temperature is raised: $Ag^+ + e^- \rightarrow Ag^0$; $Ag^0 + Ag^+ \rightarrow Ag_2^+$ ... which leads to nanocluster growth, and the reverse reactions: $Ag^+ + h^+ \rightarrow Ag^{2+}$ ... which favor nanocluster dissolution. The first series corresponds to a redox potential increase, the second to a redox potential decrease. This result provides evidence for a trend to detailed balance, leading to another important result concerning the proportion of precipitated metal. In the low-DED case (gamma irradiation at 300 K), a measurement after a few days at 300 K shows that at least 10% of the Ag ions are neutralized prior to annealing (an 80 K irradiation suggests that this proportion could even be 100% immediately after irradiation), but the proportion is reduced to about 1% after an anneal.

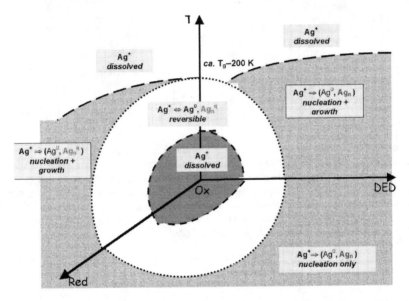

Figure 1. Schematic "phase diagram" describing Ag oligomer and nanocluster evolution in terms of temperature, ion irradiation DED and redox potential of the base host.

In the large-DED case (MeV ion irradiation at 300 K), we again find that about 10% of Ag is neutralized before annealing but in this case, annealing leads to an *increase (up to 100%)* of the Ag clustering rate. The role of the large DED is somewhat analogous to that of the developer in the photographic process, hence our introduction of the term "ion beam photography".

In summary, the ions' or photoelectrons' deposited energy density determines both the density and mobility of the electron and hole populations. The charge state of the Ag metal atom (or ion), the stability of the charge state, its diffusivity, all depend on the redox potential, hence are affected by the DED. Conversely, the DED has a major effect on the clustering efficiency. These results are summarized in the phase diagram of Figure 1, which emphasizes the similarities and differences between increasing the host's average redox potential versus increasing the ionizing radiation's DED. It also shows how combining the average host redox potential with an appropriate irradiation-induced DED allows significant control over the nanocluster nucleation and growth. This is exemplified in Figure 2, which shows cuts through two directions of the phase diagram. A detailed presentation is given in Ref. 3.

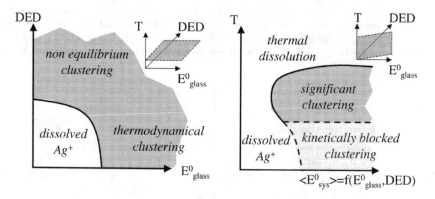

Figure 2. Cuts along two different directions of the "phase diagram" shown in Figure 1. On the left, the cross-section parallel to the basal plane illustrates how clustering may be due to (i) to the reducing potential E^0_{glass} of the glass host or (ii) via charge neutralization, to that of the ionizing radiation's DED, or (iii) to the combined action of both since they modify redox precipitation conditions. On the right, a planar cross-section containing the temperature axis illustrates the precipitation temperature-dependence when both the glass composition and the ionizing radiation DED determine the effective redox potential $<E^0_{sys}>$ (see Ref. 3).

NUCLEATION AND GROWTH OF COMPOUND SEMICONDUCTOR NANOCRYSTALS IN INSULATORS

The results of §I clearly show the influence of chemistry in nanocluster formation and growth. The alternative, of course, is to use our knowledge of chemistry. We demonstrate this using PbS nanocrystal synthesis as an example. The PbS bulk semiconductor band gap is about 0.26 eV (i.e., an emission wavelength ~3 μm), but the small carrier effective mass leads to a very large energy increase when the nanocluster size is reduced - for example, 0.83 eV (~1.5 μm) for a radius of 3 nm. Controlling the nanocluster size was therefore one of the criteria worth investigating for applications.

Sequential implantation of S and Pb in silica or glass, followed by thermal annealing, amply demonstrated [4 - 6] that attempts to synthesize compound semiconductor nanocrystals in glass or silica cannot bypass redox chemistry. Briefly, we found that implanted S appears in two valence states, one of which (-II) is trapped in a "polymer" form while the other (+VI) forms SO_3, which is very mobile. As a result, the post-annealed S profile is a composite of the implantation profile and of a very long inward diffusion profile due to the SO_3 component. Also, we observed nanocrystals of both PbS and $PbSO_4$. Since chemistry could not be bypassed, we used it [7] in the following way. (1) The host redox state (basicity) was controlled via the glass composition (i.e., pure silica was inadequate for the purpose); (2) Pb was introduced in the base glass. It has a very high solubility in glasses, which it enters in the (-II) redox state. When S is implanted, the Pb tends to bias the glass redox potential towards formation of the S(+II) state, hindering SO_3 formation and diffusion. This leads to a reasonable PbS nanocrystal size distribution (median around 3 nm, with a FWHM~25%); (3) A further redox potential bias was introduced by adding zinc oxide: Zn weakens the S(-II) bonding to glass, thus enhancing PbS formation speed and narrowing the above size distribution to a rather satisfying FWHM ~15% (Figure 3). Among the measured optical properties of such samples [8], a particularly interesting feature is the photoluminescence (PL) excitation probability, which (at 1.55 μm) is 10^5 times that of Er in silica, and 100 times that of Si nanocluster-sensitized Er in silica. The PL mechanism was shown to be - as for Si nanoclusters in SiO_2 - the bulk exciton radiative recombination as discussed in textbooks [9].

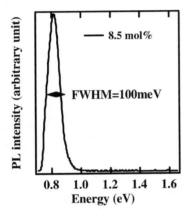

Figure 3. Photoluminescence of PbS nanocrystals (average radius 3.4 nm) synthesized by S-implantation into a glass containing 0.4 at.% Pb and 8.5 mol% ZnO (see Ref. 7).

SIGNIFICANCE OF LOGNORMAL SIZE DISTRIBUTIONS

In the course of this work, we measured many experimental nanocluster size distributions. Whereas a previous study [10] of MeV heavy-ion irradiation synthesis of Cu nanoclusters in glass had displayed a size distribution in excellent agreement with that expected from the well-known Lifshitz-Slyozov-Wagner theory [11], [12] of Ostwald ripening, we often found distributions that had shapes much closer to lognormal:

$$f_{\text{lognormal}}(R = r) = \frac{1}{r \ln \sigma \sqrt{2\pi}} \exp\left(-\frac{(\ln r/\mu)^2}{2(\ln \sigma)^2}\right)$$

where R is the cluster radius and r the running variable, μ is the geometrical average and σ the geometrical standard deviation of the distribution. Such distributions are frequently found in aerosols and in such processes as coagulation, deposition, fractal aggregation, etc., and they are by no means uncommon in nanocluster synthesis experiments, whether by ion implantation/irradiation or otherwise. We reviewed knowledge on the problem and studied [13] the origin of lognormal distributions in our cases, which provide clear test conditions. In attempting to form several compound semiconductors such as PbS, PbSe, PbTe or CdSe by "brute force" sequential implantation and annealing as discussed above, we showed that all the resulting nanocluster size distributions were lognormal, and that their so-called repartition functions

$$F_{\text{lognormal}}(R < r) = \frac{1}{2}\left[1 + \text{erf}\left(\frac{\ln r/\mu}{\ln \sigma} \right) \right]$$

were all identical (Figure 4), in spite of considerable differences in the nanocluster composition and formation conditions. Under the conditions that led to lognormal size distributions, our ex-

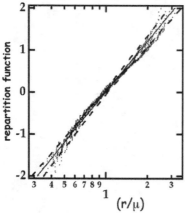

Figure 4. Radius repartition function (dots) of five different samples prepared under entirely different conditions as regards composition, concentration, concentration profile and annealing conditions. They may all be fitted by the same repartition function (full line), with $\sigma = 1.5$, to within the 5% error indicated by the dotted lines.

periments had shown that (see §II) the evolution of component profiles was complex (e.g., diffusion outside of the implant profile; correlated diffusion of the component species under annealing; chemical reactions of a component with glass host elements such as oxygen...). This contrasts with the Ostwald ripening process leading to the LSW distribution shape. In the latter instance, the system conserves the memory of its initial structure, and its evolution is uniquely determined. By contrast, the lognormal distribution is the result of processes that interfere which each other, thus progressively erasing any memory of the initial conditions.

A detailed study of the literature (see the references in [13], particularly the papers by Binder and Rosen) showed that this may be quantified. The information contained in the size distribution function f may be measured by its entropy S

$$S = -\int f(x) \ln f(x) \, dx.$$

We apply the entropy maximization principle [14]. A system's evolution is usually determined by two conservation equations: (i) matter conservation, and (ii) conservation of the size space population. The latter depends on the microscopic growth mechanism: because in the cases discussed here this mechanism is no longer defined, entropy maximization can only be performed under the sole - very general - constraint of matter conservation. This leads to an asymptotic form that is very close to the lognormal distribution function, and experiments in different areas (e. g., aggregation) have shown that the most probable geometrical standard deviation is most generally (except in some coagulation cases) in the range 1.4 - 1.5 as found here. The conclusion is that a lognormal size distribution signals a loss of memory of the nucleation and growth process, due either to the multiplication of different mechanisms or to multiple interfering processes, such as diffusion or chemical interactions with other components, that are sufficiently powerful to blur the system's memory of its initial determination.

CONCLUSIONS

The main messages of this review are that (i) because the DED affects the redox potential, combining the deposited energy density (DED) properties of ionizing particle irradiation with the host redox properties provides an effective means of controlling nucleation and growth of nanocluster assemblies; (ii) the redox properties allow sufficient control over nucleation and growth that it is possible to synthesize useful compound semiconductor nanocluster distributions; (iii) the existence of a lognormal nanocluster size distribution indicates a loss of control over the nucleation and growth process.

ACKNOWLEDGMENTS

We wish to thank our collaborators in this work: H. Béa, J. Belloni, F. Jomard, J. Majimel, J. L. Marignier, and D. Pacifici, G. Franzo, F. Priolo (University of Catania). The assistance of B. Boizot, C. Boukari and O. Kaitasov is also gratefully acknowledged.

REFERENCES

* corresponding author: bernas@csnsm.in2p3.fr
1. G. Mattei and P. Mazzoldi, Rev. Nuovo. Cim. 28(7), 1 (2005)
2. E. Cataruzza, Nucl. Inst. Meth. Phys. Res. B169, 141 (2000)
3. R. Espiau de Lamaestre, H. Béa, H. Bernas, J. Belloni and J.L. Marignier, preprint http://arXiv.org (cond-mat 0703510)
4. R. E. de Lamaestre, H. Bernas, F. Jomard and J. Majimel, J. Non-Cryst. Sol. 351, 3031 (2005)
5. R. E. de Lamaestre, F. Jomard, J. Majimel and H. Bernas, J. Phys. Chem. B109,19148 (2005)
6. R. E. de Lamaestre, H. Bernas, C. Ricolleau and F. Jomard, Nucl. Inst. Meth. Phys. Res. B242, 214 (2006)
7. R. E. de Lamaestre and H . Bernas, J. Appl. Phys. 98, 104310 (2005)
8. R. E. de Lamaestre, H. Bernas, D. Pacifici, G. Franzo and F. Priolo, Appl. Phys. Lett. 88, 181115 (2006)
9. S. V. Gaponenko, Optical Properties of Semiconductor Nanocrystals, Cambridge Univ. Press, Cambridge, UK (1998)

10. E. Valentin, H. Bernas, C. Ricolleau and F. Creuzet, Phys. Rev. Lett. **86,** 99 (2001)
11. I. M. Lifshitz and V. V. Slyosov, J. Phys. Chem. Solids **28,** 35 (1961)
12. C. Wagner, Z. Elektrochem. **65,** 58 (1961)
13. R. E. de Lamaëstre and H. Bernas, Phys. Rev. **B73,** 125317 (2006) and refs. therein
14. J. M. Rosen, J. Colloid Interface Sci. **99,** 9 (1984)

Mater. Res. Soc. Symp. Proc. Vol. 1020 © 2007 Materials Research Society   1020-GG06-04

Controlled Growth of Conducting Carbon Nanowires by Ion Irradiation: Electrical and Field Emission Properties

Amit Kumar[1], L. D. Filip[2], J. D. Carey[2], J. C. Pivin[3], A. Tripathi[1], and D. K. Avasthi[1]

[1]Materials Science Group, Inter-University Accelerator Centre, Aruna Asaf Ali Marg, PB-10502, New Delhi, 110067, India
[2]ATI, University of Surrey, Guildford, GU2 7XH, United Kingdom
[3]CSNSM, Orsay, 91405, France

ABSTRACT

The conducting carbon nanowires embedded in fullerene matrix are synthesized by high energy heavy ion irradiation of thin fullerene film. We report the control growth of carbon nanowires, their electrical and field emission properties. The typical diameter of the conducting tracks is observed to be about 40 to 100 nm. The conductivity of the conducting zone is about seven orders of magnitude higher than that of the fullerene matrix. The conducting nano wires are evidenced by conducting atomic force microscopy. All the nanowires are parallel to each other and are perpendicular to the substrate. The density (spacing), growth direction and length of these carbon nanowires simply can be changed by ion fluence, angle of irradiation and the film thickness, respectively. The field emission measurements on these nanowires reveal that the threshold voltage is about (~13 V/μm), whereas the as deposited fullerene films shows a break down at (~ 51 V/μm). The present approach of making controlled conducting carbon nanowires is quite promising, as it takes few seconds of ion irradiation and no catalyst is required.

INTRODUCTION

The carbon-based materials are expected to be key materials in the future electronics because of their unique characteristics and biocompatibility. The construction of nano-electronic devices is technologically challenging and many different approaches are possible. The controlled growth of a single nanowire or an ensemble of nanowires, their growth direction, suitable alignment and spacing on a substrate is of importance for the applications. The simplification of the nanowire fabrication procedure as well as the development of large scale and low price production methods remains open problems. Ion beam methods are now emerging as a tool in nano-fabrication for structuring material and modification on the nanometer/atomic scale [1-3].

The present paper reports the formation of conducting carbon nanowires by heavy ion irradiation in fullerene films. The effect of the density of energy transferred by ions to target electrons on the dimension and electrical properties of the nanowires are discussed. Their field emission properties are also studied.

EXPERIMENTAL DETAILS

Fullerene thin films of 200 nm thicknesses were deposited on 50 nm thick Au layers on Si substrate. Gold films of 50 nm in thickness were deposited on Si substrate by resistive heating. Subsequently, the fullerene films were deposited on these Au films by sublimation of a C_{60} pellet. A small area of the Au film was masked so that it can be used as a contact for conducting AFM measurement. The X-Ray diffraction pattern of the as-deposited films revealed the polycrystalline nature. Raman measurements on as-deposited film show the absence of amorphous carbon phase. The fullerene films were irradiated by 120 MeV Au, 180 MeV Ag and 55 MeV Ti ions at different fluences using 15MV Pelletron accelerator at Inter-University Accelerator Centre, New Delhi. The energy loss and range of the ion estimated by Monte-Carlo simulation package (SRIM 2003) [5], given in table I. The conducting AFM (C-AFM) measurements were performed on the pristine and irradiated samples using Nanoscope IIIa SPM. The field emission (FE) characteristics of pristine and irradiated films were examined using a sphere-to-plane geometry in which a 5 mm stainless-steel ball bearing is suspended 40 mm above the surface at a high positive potential in a vacuum better than 4×10^{-6} mbar. The threshold field (E_{th}) is defined as the macroscopic electric field, which gives an emission current of 1 nA for a conditioned film.

Table I. The ion mass, energy, electronic energy deposition, range of ions in fullerene matrix and electrical conductivity of synthesized carbon nanowires.

Ion and energy	Electronic energy loss ((keV/nm)	Ion Range (Om)	Conductivity (S/cm)
55 MeV $_{22}$Ti^{48}	5.6	13.6	10^{-4}
180 MeV $_{47}$Ag^{107}	11.8	22.8	10^{-2}
120 MeV $_{79}$Au^{197}	13.7	16.6	10^{-1}

RESULT AND DISCUSSION

The conducting AFM image of the fullerene film irradiated for 120 MeV Au ions at a fluence of 6×10^{10} ions/cm^2 are shown in figures 1 (a). The Z-axis of figure represents the current. It can be clearly seen that the current in ion tracks is significantly higher than that of the region not hit by the ions. The sectional analysis reveals that the diameter of conducting zone varies from 40 nm to 100 nm. The observed tracks show some mismatching to incident ions. Mismatch between fluence and the ion track areal density may result from random ion impacts resulting in overlapping tracks and/or lateral conduction with the neighboring tracks. In addition, the finite size of tip may have influence. Analyses of irradiated films by Fourier Transform Infra Red spectroscopy indicate the simultaneous formation of polymerized C_{60} and amorphous carbon. It has been proposed that these transformation occur respectively in annular cylindrical regions (called track halo) perturbed essentially by secondary electrons and in track cores where high densities of electronic excitations and ionizations are produced directly by ions [4]. These results demonstrate that conducting carbon nanowires, which are perfectly perpendicular to the surface, and all the conducting channels are parallel to each other.

The length of the nanowires can be tuned by the thickness of fullerene film and its orientation (vertical alignment) can be engineered simply changing the incidence angle of the ion beam. The inter-nanowire spacing can be controlled to some extent by the ion fluence. No catalysts are required for growth and only a few seconds are required for nanowire growth.

The role of electronic density deposition on the nanowires has been investigated by irradiation of same thickness fullerene films by 120 MeV Au (S_e = 13keV/nm), 180 MeV Ag (S_e = 11keV/nm) and 55 MeV Ti (S_e = 5 keV/nm) ions. The conducting paths (nanowires) are observed in the above irradiated films at 0.5 V, 2V and 9 V corresponding to the electronic energy deposition of 13 keV/nm, 11 keV/nm and 5 keV/nm, details are given in table I. These values show that electronic energy deposition played main role in conducting track formation. It is noticed that in case of 120 MeV Au ions irradiated film, the track diameter is larger than 180 MeV Ag ion irradiated film. The larger diameter in case of Au ion is due to larger electronic energy deposition. In case of 55 MeV Ti ion irradiated films, we observed conducting track at 9 V. At 9 V, the tracks diameter is bigger and seem to pull-up from surrounding the ion tracks, it means the currents are coming from the track surrounding at such high voltage. It is also noticed that for Ti ions, the mismatch between incident ions and observed tracks is quite higher (as shown in figure 3 c) than Au and Ag ions. The observations of (i) conducting track at higher voltage and (ii) higher mismatch between the incident ions and observed track indicate the possibility of the discontinuous tracks.

The conductivity of the nanowires for 180 MeV Ag ion irradiated films is about 0. 03 S/cm, for the track diameter of 45 nm and 46 nA current corresponding to 2 V applied bias. The highest conductivity of the conducting zone is about 0.1 S/cm for 120 MeV Au ion, which is almost seven orders of magnitude higher than the conductivity of the pristine fullerene films (~ 10^{-8}S/cm) [6]. The conductivity of these nano wires varies from ~ 10^{-1} S/cm to ~ 10^{-4} S/cm, depending on Se.

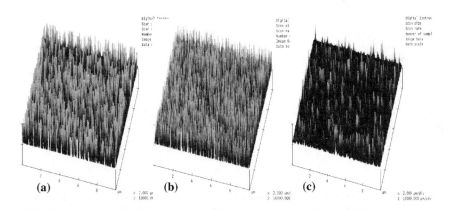

Figure 1. (a), (b) and (c) conducting AFM image of the fullerene films (~200 nm) irradiated by 120 MeV Au, 180 MeV Ag and 55 MeV Ti ions respectively at the same

fluence of 6×10^{10} ions/cm^2. The vertical nanowires represent the current flowing through the conducting ion tracks.

Field emission properties have been studied on pristine and synthesized carbon nanowires by 180 MeV Ag ions at different fluences, as shown in figure 2 (a). The threshold voltage of about 13, 15, 17 V/μm has been observed for carbon nanowires synthesized at 3×10^{11}, 3×10^{12} and 3×10^{13} ions/cm^2 fluence respectively, whereas pristine films show breakdown behavior at 51 V/μm (shown in the inset of figure 2 (a)). The Fowler – Nordheim (F-N) plot [7,8] for field emission current of figure 2(a), $\ln(I/V^2)$ versus $1/V$ given in figure 2 (b). The inset of figure 2 (b) describe the simple form of the Fowler-Nordheim equation.

The linear dependence of the plot indicates that the observed filed emission from the carbon nanowires follow tunneling mechanism of fowler – Nordheim equation. The slope of the F-N plot depends on the work function, the field amplification factor and inter electrode distance. The slope of the F-N plot can be given by, - b $(\phi^{3/2})/\beta$, where b is a constant with the value of 6.83×10^{-9} eV$^{-3/2}$ V/micron, ϕ is work function and β is field amplification/enhancement factor. We estimated the field enhancement factor (β) value with the hypothesis that the work function is similar to that of graphite ($\sim$ 5 eV) and found $\beta = 447$, 343 and 291 at 3×10^{11}, 3×10^{12} and 3×10^{13} ions/cm^2 fluence, respectively.

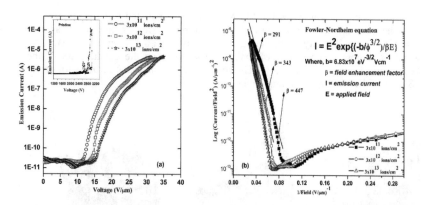

Figure 2. (a) The emission current versus applied voltage of the synthesized carbon nanowires by ion irradiation. Inset shows the emission from the pristine fullerene film. (b) The Fowler-Nordheim equation fitting of I-V measurement of the synthesized carbon nanowires.

These analyses show that field enhancement factor increases as a function of ion fluence. It is reported that at higher areal densities of carbon nanotubes field screening becomes a significant factor for emission properties, which leads to lower filed enhancement factor [9]. In present case, lower field enhancement factor at higher fluence is due to the less inter-spacing between the nanowires or higher areal density. Similar to present work, Schwen et al [10] also reported that the larger threshold applied fields of 30 V/μm required for 350 MeV Au ion irradiated tetrahedral amorphous carbon (a-C) films, grown by mass selected ion beam deposition. They attributed the high threshold field solely to the embedding of the nanowires in the a-C matrix whose dielectric constant reduces the field experienced the nanowire and consequentially larger applied fields are required for emission.

CONCLUSION

Energetic hevay ions, having a straight path in materials and producing high densities of electronic excitations, are very efficient for synthesizing conducting carbon nanowires in fullerene thin films. No catalyst or any further purification is required in present ion beam based method, unlike in the case of carbon nanotubes by chemical routes. Irradiation using the mask of suitable geometry can be used to create nanowires as desired. Conductive AFM measurements show indicates that the continuity of conducting paths increases with the density of electronic energy loss of ions. The threshold voltage of field emission measured for these nanowires is relatively low and the emission obeys the tunneling mechanism by Fowler- Nordheim equation. Irradiation only requires a short time, which allows for synthesis in large quantities. Thus the present approach of making carbon nanowires is quite promising and these nanowires may be useful for better field emitters and large scale applications.

ACKNOWLEDGMENTS

A.K. is thankful to Council of Scientific and Industrial Research (CSIR), New Delhi, India for providing fellowship. The authors (DKA and JCP) are thankful to the Indo French Centre for Promotion of Advanced Research for financial support to carry out work on 'Generation of nano phase by energetic ion beams'. We are also thankful to DST for providing the financial support to procure AFM/C-AFM under the IRPHA project. Work at ATI Surrey is sponsored by the EPSRC.

REFERENCES

1. Amit Kumar, D.K.Avasthi, J.C. Pivin, A. Tripathi, and F. Singh, *Physical Review B.* 74, 153409 (2006).
2. Amit Kumar, Fouran Singh, J, C. Pvin, *Journal of Physics D: Applied Physics 40,* 2083 (2007).
3. M. Toulemonde, J. M. Costantini, Ch. Dufour, A Meftah, E. Paumier, and F. Studer, *Nucl. Instr. and Meth. B* 116, 37 (1996).
4. Amit Kumar, D.K. Avasthi, A. Tripathi, D. Kabiraj, F. Singh, J.C. Pivin, *Journal of Applied Physics* 101, 014308 (2007).
5. J. F. Zeigler, J. P. Biersack, V. Littmark,."*The Stopping and Range of Ions in Solids*" Pergamon, New York (1985).
6. Amit Kumar, F. Singh, R. Kumar, J. C. Pivin, D.K. Avasthi, *Solid state communication* 138, 448 (2006).

7. Toshiki Hara, Jun Onoe, and Kazuo Takeuchi, *J. Appl. Phys.* 92, 7302 (2002).
8. J. D. Carey, R. D. Forrest, R. U. A. Khan and S R P Silva, *Appl. Phys. Lett.* 77, 2006 (2000).
9. L. Nilsson, O. Groening, C. Emmenegger, O. Kuettel, E. Schaller, L.Schlapbach, H. Kind, J.-M. Bonard, and K. Kern, *Appl. Phys. Lett.* 76, 2071 (2000).
10. D. Schwen, C. Ronning and H. Hofsass, *Dianond and Related Materials* 13, 1032 (2006).

Mater. Res. Soc. Symp. Proc. Vol. 1020 © 2007 Materials Research Society 1020-GG06-06

Nanofabrication Based on Ion Beam-Laser Interactions with Self-Assembly of Nanoparticles

N. Kishimoto[1], K. Saito[1], Jin Pan[2], H. Wang[1], and Y. Takeda[1]
[1]Quantum Beam Center, National Institute for Materials Science, 3-13 Sakura, Tsukuba, 305-0003, Japan
[2]Univ. of Tsukuba, Tsukuba, Japan

ABSTRACT

Ion beam-based techniques offer various possibilities for robust spatial control of nanoparticles. Since ion implantation is inherently good at depth control of solutes or nanoparticles, additional lateral control may lead to 3D control of nanoparticles. We pursue a lateral-control method of nanoparticle assembly by controlling photon-energy field under ion implantation. Laser is irradiated into a-SiO$_2$, either sequentially or simultaneously with ion implantation. Ions of 60 keV Cu$^-$ or 3 MeV Cu^{2+} and photons of 532 nm are used to study effects on nanoparticle evolution. Simultaneous laser irradiation under ion implantation enhances surface plasmon resonance (SPR), i.e., nanoparticle precipitation, while sequential laser irradiation of 532 nm tends to cause a decay of SPR, i.e., dissolution of Cu nanoparticles. The energy-field perturbation of laser, interactive with nanoparticle evolution, can be used for controlling nanoparticle assembly.

INTRODUCTION

Ion beam-based techniques offer various possibilities for robust spatial control of nanostructures, either in a self-assembling- [1] or in a controlled manner [2, 3]. In order to accomplish the ion-beamed-based nanofabrication for industrial applications, one of the most important targets is 3D spatial control of nanostructures or nano-doping, utilizing characteristic advantages of ion beam methods. Towards the 3D spatial control of nanostructures, we have to carry out not only the instrumental development with a high resolution less than a few nm, but also materials scientific research on in-beam- [4] or post-implantation kinetic processes.

Among various possible applications, metal nanoparticles embedded in insulators are one of the most attractive options, since the ion implantation technique is originally suitable for injection of immiscible metal ions into transparent substrates. Moreover, recent studies have revealed fascinating plasmonic functions due to the surface plasmon resonance [5], such as ultrafast nonlinear devices [6], near-field optical waveguides [7], biosensors [8] and so on.

Since ion implantation is inherently good at depth control of solutes or nanoparticles, additional lateral control should result in 3D control of nanoparticles or nano-doping. In principle, two kinds of approach for the lateral control are possible: one is patterning the ion/atom supply (supply control) such as masked implantation [9], FIB [10], Ion Projection Lithography (IPL) [2,3], etc. and the other is patterning interactive fields with solutes/precipitate states (perturbation control) such as photon, phonon fields, mechanical stress fields, etc. For instance, either photon-energy perturbation [11] or nano-indentation-induced perturbation [12] tends to cause precipitation or redistribution of solutes. Although the supply-patterning methods are obviously more authentic ways to form the patterned nanostructures, the perturbation methods may hold as a complementary way, having advantages of low cost, easiness in handling and, more importantly, freedom from surface sputtering (as will be discussed later). The phase control of precipitates versus solutes is acceptable in the present case, since the plasmonic functions originate solely from the nanoparticle phase, not from the solute states [6, 13]. To control nanoparticle assemblies, self-assembling mechanisms have to also be handled, even by IPL. In this paper, we present how to control nanoparticle assemblies by controlling energy fields with laser irradiation under ion implantation, and discuss tactics towards 3D control of nanoparticles.

EXPERIMENTAL PROCEDURES

Negative Cu ions of 60 keV or Cu^{2+} ions of 3 MeV were used for ion implantation into amorphous (a-) SiO_2 (KU-1: 820 ppm OH-), to change the depth of nanoparticles. Pulsed laser of 2.3 eV (532 nm, Nd: YAG with second harmonic generator) was used to add photon-energy field, under or after the ion irradiation. Usage of negative ions is important for the lower-energy implantation into insulators, especially when studying surface interactions, since the negative ions automatically alleviate surface charging balancing with secondary electron emission [14]. It can be pointed out that the space charge or surface-charging may become a critical issue when ion beam pattern goes down to a nano-scale. The projected ranges of 60 keV or 3 MeV Cu ions were estimated by the SRIM code [15] to be about 50 nm or 1.8 μm, respectively. The ion flux and the laser intensity were 2-10 μA/cm² and 0.001 - 0.3 J/cm²-pulse (pulse-width of 7 nsec at 10 Hz). The ion fluence was varied from 1×10^{16} to 3×10^{17} ions/cm². Simultaneous irradiation of Cu ions and 532 nm laser was carried for the a-SiO_2 substrates, as well as the single ion irradiation. After the ion irradiation, optical absorption (photon energy: 0.5 to 6.5 eV) and cross-sectional TEM were conducted to study correlation between surface-plasmon resonance and nanoparticle morphology.

For comparison, masked ion-implantation with a patterned resist mask was also conducted for 60 keV Cu ions. A patterned mask of photoresist AZ3100 was formed by laser-drawing lithography with a resolution of 0.5 μm. A 1D grating mask with a pitch of 2 μm was firstly fabricated on the surface, and 60 keV Cu- ions were irradiated. After removing the resist mask, the second grating in the orthogonal direction was formed and the ion irradiation was made to achieve the 2D-lattice structure.

RESULTS AND DISCUSSION

Towards the 3D control of nanoparticles, the issues of 1D control are briefly discussed and move to those of additional 2D control by using laser irradiation under heavy ion irradiation. Subsequently, the result of conventional masked implantation is discussed in comparison with the laser irradiation method.

1D control of nanoparticles

Though ion implantation is basically favorable for 1D control along with depth of a substrate, the stochastic nature of ion-implantation processes restricts the spatial resolution in depth. For instance, the straggling blur of Cu ions is ±250 nm with a projected range of 1.8 μm in SiO_2. As far as ion/atom collision and scattering are concerned, this stochastic blurring is physically unavoidable. If thermal annealing is applied to a Cu-implanted specimen in order to form metal nanoparticles, the thermal diffusion of implants causes even larger blurring [16]. Since metal nanoparticle size suitable for plasmonics is in the order of 10 nm, the stochastic blurring appears to be significant.

This kind of problems is in common between the supply- and the perturbation controlling methods mentioned above. There are no such concerns necessary for ion-beam fabrication methods which use ion-beam energy at the beam-entry surface, e.g., nanofabrication of photoresist [2,3], FIB-induced nanofabrication [10] and magnetization reorientation of CoPt [17]. However, once impurity doping or nanoparticle formation is aimed at, the spatial resolution may be eventually limited by the ion straggling.

One of the methods to alleviate this constraint is to use the lower energy: Usage of 60 keV Cu ions gives a straggling of ± 15 nm, which will be mostly acceptable. Another method is to utilize self-assembling phenomenon [18].

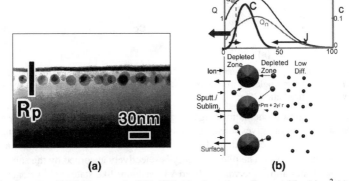

(a)   (b)

Fig. 1 Cross-sectional TEM image of a 60 keV Cu$^-$-implanted specimen at 45 $\mu A/cm^2$ (a) and the schematic diagram of the mechanism (b). The upper-right is implant distribution C, energy distributions of Q_e (electronic stopping) and Q_n (nuclear energy deposition). The lower-right is an illustration of the self-assembly of nanoparticles.

Figs. 1 (a) and (b) show a cross-sectional TEM image of a Cu$^-$-implanted specimen at an appropriate ion flux (45 $\mu A/cm^2$) and the schematic diagram of the mechanism, respectively. As ion flux increases, a majority of the nanoparticles are coarsened and results in narrowing of the depth profile [18]. As shown in Fig. 1(a), the nanospheres are linearly positioned along a certain depth shallower than the projected range R_P. In this case, the ion beam not only causes spontaneous precipitation of Cu implants, but also the localized gradient energy collects Cu solutes into a certain depth, together with the Ostwald-ripening mechanism. This self-assembling mechanism provides 1D control of nanoparticles along the depth, possibly better than the physical constraint of the straggling. Similar narrowing of nanoparticle distribution was found in Cu-implanted PMMA [19] and seems to be a fairly universal phenomenon. Accordingly, the first step in the 3D-control scenario is the 1D control with ion-energy induced self-assembly of nanoparticles. It is noted that the ion-beam energy plays an important role in the self-assembly.

2D lateral control of nanoparticles

After the 1D depth control of nanoparticles is secured, the second necessary step is the 2D lateral control of nanoparticles. As was indicated by the flux-dependent self-assembly of nanoparticles, electronic excitation may influence the nanoparticle formation in the substrate [11]. To attempt nanoparticle control with the electronic excitation, desirable is the external perturbation which is independent of the ion beam and energy-selective. Accordingly, we apply laser irradiation of a given photon energy to give pure electronic excitation, either simultaneously or sequentially with ion implantation.

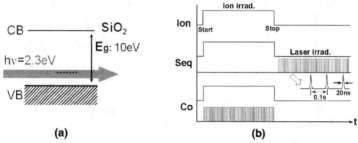

(a)   **(b)**

Fig. 2 Schematic energy diagram (a) and time sequences of combined irradiation of ion and laser (b). The upper is single ion irradiation (Ion), the middle sequential irradiation (Seq) and the lower is co-irradiation (Co).

Fig.2 (a) shows schematic energy diagram of SiO_2 including laser energy of 2.3 eV. Selection of laser energy gives selective behaviors of electronic states. Unless sub-gap states (defects, solutes, etc) are created, the laser of 2.3 eV (532 nm) is transparent for a SiO_2 substrate with a wide gap (~10 eV). The photon energy is selectively absorbed by the sub-gap states (acceptor-like states above the valence band VB or donor-like states below conduction band CB), whereas electronic stopping of ions has no such selectivity. The selectivity of excitation is one of the most important characteristics in laser irradiation. The time sequences of laser irradiation are given in Fig. 2(b). To distinguish dynamic effects from the total-energy-driven effects, both co-irradiation and sequential irradiation were conducted for comparison, keeping the same total energy. If any changes between the two modes of laser irradiation were observed, the difference should be attributable to in-beam dynamic energy absorption.

Atomic injection by the ion implantation method is a non-equilibrium process, which seems to be similar to rapid solidification of a melted solution. Since the laser interaction does not change the total amount of implants for a given ion fluence, distinct phase control between solutes and precipitates is desirable and may be realized in the vicinity of a supersaturated condition of solutes.

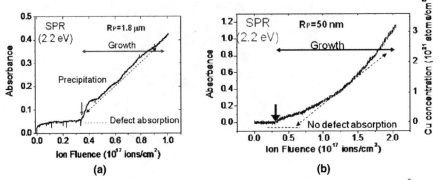

Fig. 3 Ion fluence dependence of SPR absorbance under single ion irradiation of 3 MeV Cu^{2+} (a) and 60 keV Cu^- (b).

solution. Figs. 3(a) and (b) show in-beam monitoring of optical absorbance at the photon energy corresponding to surface plasmon resonance. The ion-fluence dependence of optical absorbance at 2.2 eV shows a threshold-type variation above a certain fluence of 3×10^{16} ions/cm^2. Above this critical fluence, the absorbance rapidly increases, whereas, below the critical fluence, the absorbance keeps a small level irrespective of the ion fluence. The threshold-type variation clearly indicates presence of supersaturated solid-solution and that single ion irradiation induces spontaneous precipitation with the ion energy above the critical fluence. The critical fluence corresponds to peak concentrations of 1 and 10 at.% for 3 MeV and 60 keV, respectively (after the SRIM simulation [15]), both of which are much higher than the thermodynamic solubility of Cu in SiO_2, i.e., 0.1 ppb [20]. Accordingly, the implanted state is primarily in the supersaturated condition, and the critical fluence is effectively determined by the balance between the beam energy (ion flux) and the local concentration of Cu implants.

Consequently, we may utilize the supersaturated state around or above the critical fluence, in order to obtain contrastive change of precipitates from solutes. Fig. 4 shows cross-sectional TEM images of a-SiO_2 that was ion-alone irradiated (a), sequentially irradiated (b) and co-irradiated (c) with 3 MeV Cu^{2+} ions (10 µA/cm^2) and 532 nm laser (0.2 J/cm^{2}pulse), where the total ion dose is 3×10^{16} ions/cm^2 and the imaging area corresponds to the vicinity of the projected range.

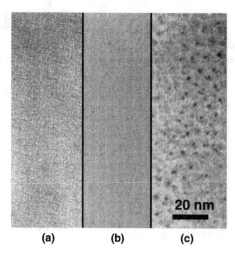

(a)　　　　(b)　　　　(c)

Fig. 4 Cross-sectional TEM images of a-SiO$_2$ that were single-ion-irradiated (a), sequentially irradiated with 3 MeV Cu^{2+} and the 532 nm-laser (b) and co-irradiated with the Cu ions and the 532 nm-laser (c), to the total ion dose of 3×10^{16} ions/cm^2, where the imaging areas correspond to the vicinity of the projected range.

Although the ion fluence is the same among the three irradiation modes, the microstructure of co-irradiation (c) is significantly different from the other (a) and (b). Only when ion and laser are simultaneously irradiated (Fig. 4(c)), finely dispersed Cu precipitates emerge in the a-SiO$_2$ matrix. On the other hand, either the single ion irradiation (Fig. 4(a)) or the sequential irradiation (Fig. 4 (b)) gives rise to no precipitation.

In the sequential irradiation mode, the laser irradiation begins after the maximum accumulation of solutes or residual defects. The sequential irradiation in this study is similar to post-implantation laser annealing. In general, there is abundant evidence of laser-induced thermal annealing/precipitation in semiconductors and insulators. However, thermally-induced laser annealing does not take place at this ion fluence, as shown in Fig. 4(b). Since the enhanced precipitation is observed only for the co-irradiation condition, it is judged that energy absorption by solutes is not responsible for the precipitation. It is thus suggested that the co-irradiation-induced precipitation may primarily be an athermal process around this fluence and that the energy absorption for precipitation is caused by anything dynamical under the ion implantation. In fact, it has been revealed by in-situ measurement of optical absorption spectra [4] that transient defect states (~self trapped holes [11, 21]), coinciding with the laser energy of 2.3 eV, are created by the ion irradiation and act as efficient energy absorbers of the laser. The laser energy absorbed leads to annealing of electronic defects followed by precipitation. It should be noted that this athermal feature does not always hold. If the ion fluence becomes much higher, the implanted region tends to be metallic and nano-second laser may give rise to thermal heating, more or less. The result of Fig. 4 indicates that the laser co-irradiation under ion implantation can accomplish phase control between precipitates and solutes, as far as the irradiation condition (mostly ion fluence) is optimized.

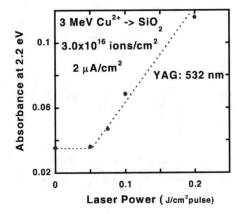

Fig. 5 Laser-power dependence of SPR absorbance for a-SiO$_2$ co-irradiated with 532 nm-laser and 3 MeV Cu^{2+} to 3×10^{16} ions/cm^2.

To attain contrastive phase control, dependence of nanoparticle precipitation on laser intensity is important as well as the photon-energy selectivity and the threshold-type fluence-dependence mentioned above. Fig. 5 shows laser-power dependence of the absorbance at SPR energy of 2.2 eV for a-SiO$_2$ co-irradiated with 532 nm-laser and 3 MeV Cu^{2+}. The laser power dependence of the SPR shows threshold-type dependence above 0.05 J/cm^{2}pulse. This behavior also helps obtaining contrastive phase control with the laser co-irradiation method.

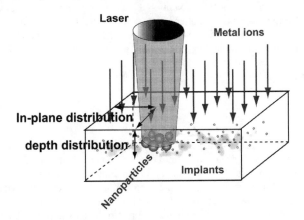

Fig. 6 Schematic diagram of ion-laser combined irradiation method for nanoparticle control.

Fig. 6 shows a schematic diagram of the ion-laser combined irradiation method. The laser energy can be given by either laser drawing or laser patterning. The requirements of laser-ion coincidence and of threshold-type behaviors are the keys for the lateral control of nanoparticles with the photon-energy perturbation method. In addition, it should be pointed out that too much laser power for well-precipitated specimens tends to give rise to dissolution of pre-existent nanoparticles [23] and it is also usable for erasing nanoparticles.

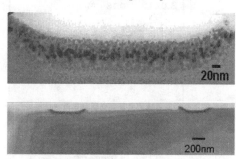

Fig. 7 Cross-sectional TEM images of a-SiO$_2$ that was implanted with 60 keV Cu⁻ to 4×10^{16} ions/cm^2 by conventional masked implantation. The upper image is the enlargement of the lower one.

Finally let us discuss problems of conventional masked implantation. Fig. 7 shows cross-sectional TEM images of a-SiO$_2$ implanted beyond the critical fluence (Fig. 3(b)), by using a micro-patterned photoresist mask (AZ3100). As seen in the lower magnification image (the lower), nanoparticle zones of 500 nm wide at a period of and 2 μm are successfully patterned [24]. However, the enlarged image (the upper) shows occurrence of pronounced sputtering and non-uniform distribution of nanoparticles towards the boundary. The surface recession and the proximity effects are inevitable for the conventional masked implantation. Though the lower energy (shallower) implantation is the better to attain depth-directional control of nanoparticles, there is the lower-energy limit to obtain well-defined buried nanostructures. The sputtering-related constraint is universal among ion/atom supply methods including IPL. On the other hand, the perturbation methods are almost free from the constraint, though the lateral resolution is limited to a submicron level.

CONCLUSIONS

Among various ion beam-based techniques to fabricate nanostructures, nanoparticle control is an attractive target for plasmonic applications, but essential difficulty arises from the nature of the secondary process after the primary atomic injection. Since the secondary process is possibly influenced by the electronic-energy perturbation, the ion-laser combined irradiation has been studied aiming at self-assembling lateral-control of nanoparticles, towards the 3D control. The lateral control method is regarded as controlling photon-energy perturbation under ion implantation. One of the most important aspects for nanoparticle spatial control is to understand and utilize energy-driven kinetic processes as well as nano-patterning of the ion/atom supply or the energy perturbation.

The 1D depth-control of nanoparticles is automatically guaranteed, to some extent (within the ion-straggling width), as far as the ion implantation methods are applied. To surpass the straggling limit, self-assembling alignment of nanoparticle may be applicable if the in-beam kinetic processes are optimized. Usage of the ion-beam (electronic) energy is a key to control spontaneous precipitation and to sharpen nanoparticle distribution.

The additional 2D-lateral control of nanoparticles is attainable by the ion-laser combined-irradiation method. Ions of 60 keV Cu^- or 3 MeV Cu^{2+} and photons of 532 nm were used to study effects on nanoparticle evolution. It has been demonstrated that simultaneous laser irradiation under ion implantation significantly enhances nanoparticle precipitation, while sequential laser irradiation tends to dissolve pre-existent nanoparticles. The important factors in controlling nanoparticle assemblies are usage of supersaturated solid-solution conditions, coincidence of laser and ion, threshold-type laser-power dependence, selectivity of photon energy and other kinetic features. Since the SPR-related properties arise only from the precipitate phase, only the phase control should work out without patterning the solute distribution.

Thus, the energy-field perturbation of laser, interactive with nanoparticle evolution, is applicable for controlling nanoparticle assembly, with considerable easiness, low cost and probably high throughput. The kinetic aspects of nanostructural control are important in common with other ion/atom supply-controlling methods.

Further optimization of kinetic processes and the actual patterning are necessary in order to demonstrate the feasibility of nanoparticle patterning.

ACKNOWLEDGMENTS

A part of this study is financially supported by the Budget for Nuclear Research of Ministry of Education, Culture, Sports, Science and Technology, Japan, based on screening and counseling by the Atomic Energy Commission. The authors would like to sincerely thank Dr. John Baglin, IBM Almaden Research Center and Prof. Daryush Ila, Alabama A&M University for fruitful discussion and collaboration for ion-beam nanofabrication.

REFERENCES

1. For instance, A.D. Brown, H.B .George, M.J. Aziz and J.D. Erlebacher, Materials Research Society Symposium Proceedings **792** , R7.8 (2004).
2. H. Loeschner, G. Stengl, R. Kaesmaier and A. Wolter, J. Vac. Sci. Technol. B 19 (2001) 2520.
3. X. Jiang, Q. Ji, L. Ji, A. Chang and K-N. Leung, J. Vac. Sci. Technol. B 21 (2003) 2724.
4. O.A. Plaksin, Y. Takeda, H. Amekura and N. Kishimoto, J. Appl. Phys. 99 (2006) 044307.
5. F. Hache, D. Ricard, C. Flytzanis, J. Opt. Soc. Am. B 3 (1986) 1647.
6. Y. Takeda, O. Plaksin, L. Lu and N. Kishimoto, Nucl. Instrum. & Meth. in Phys. Res. B242 (2006) 194.
7. S. A. Maier, M.L. Brongersma, P.G. Kik, S. Meltzer, A.A.G. Requicha and H.A. Atwater, Adv. Mater. 13 (2001) 1501.
8. J. Dostalek, J. Jiang, J. Ladd, Surface Plasmon Resonance Based Sensors, Springer Series on Chemical Sensors and Biosensors, Springer (2006) .

9. R. Fromknecht, G. Linker, K. Sun, S. Zhu, L.M. Wang, A. van Veen, M.A. van Huis, T. Weimann, J. Wang, J. Niemeyer, F. Eichhorn and T. Wang, Mat. Res. Soc. Symp. Proc. 792 (2004) R8.3.1.

10. T. Morita, K. Kanda, Y. Haruyama and S. Matsui, Japanese. J.Appl. Phys., 44(2005) 3341.

11. N. Kishimoto, O.A. Plaksin, K. Masuo, N. Okubo, N. Umeda and Y. Takeda, Nucl. Instrum. & Meth. in Phys. Res. B242(2006) 186.

12. J. Pan, H. Wang, Y. Takeda, N. Umeda, K. Kono, H. Amekura and N. Kishimoto, Nucl. Instrum. & Meth. B257 (2007) 585.

13. Y. Takeda, O.A. Plaksin, J. Lu and N. Kishimoto, Vacuum 80 (2006) 776.

14. Ishikawa, J, Tsuji, H, Toyota, Y., Gotoh, Y., Matsuda, K., Tanjyo, M. and Sakaki, S., Nucl. Instrum. & Meth. in Phys. Res. B96, 7 (1995).

15. J.F. Ziegler, J.P. Biersack and U. Littmark, The Stopping and Range of Ions in Solids, (Pergamon Press, New York, 1985), Chap 8.

16. N. Umeda, N. Kishimoto, Y. Takeda and C.G. Lee, Nucl. Instrum. & Meth. B (2000) 864.

17. D. Weller, J.E.E. Baglin, A.J. Kellock, K.A. Hannibal, M.F. Toney, G. Kusiski, S. Lang, L. Folks, M.E. Best and B.D. Terris, J. Appl. Phys., 87 (2000) 5768.

18. N. Kishimoto, N. Umeda, Y. Takeda, C.G. Lee, and V.T. Gritsyna, Nucl. Instrum. & Meth, in Phys. Res. B 148 (1999) 1017.

19. H. Boldyryeva, N. Umeda, Oleg Plaksin, Y. Takeda and N. Kishimoto, Surf. & Coat. Tech., 196 (2005) 373.

20. J.D. McBrayer, R.M. Swanson and T.W. Sigmon, J. Electrochem. Soc., 133 (1986) 1242.

21. Y. Sasajima and K. Tanimura, Phys. Rev. B68 (2003) 14204.

22. K. Masuo, O.A. Plaksin, Y. Fudamoto, N. Okubo, Y. Takeda, N. Kishimoto, Nucl. Instrum. & Meth. in Phys. Res. B 247 (2006) 268–270.

23. H. Wang, Y. Takeda, N. Umeda, K. Kono and N. Kishimoto, Nucl. Instrum. & Meth. in Phys. Res. B257 (2006) 20.

Mater. Res. Soc. Symp. Proc. Vol. 1020 © 2007 Materials Research Society   1020-GG06-09

Core-Satellite Metallic Nanoclusters in Silica Obtained by Multiple Ion Beam Processing

Giovanni Mattei[1], Valentina Bello[1], Paolo Mazzoldi[1], Giovanni Pellegrini[1], Chiara Maurizio[2], and Giancarlo Battaglin[3]

[1]Department of Physics, University of Padova, via Marzolo 8, Padova, I-35131, Italy
[2]ESRF, GILDA-CRG, CNR-INFM, Rue Horowitz 6, B.P. 220, Grenoble, 38043, France
[3]Department of Physical Chemistry, University of Venice, Dorsoduro 2137, Venice, I-30123, Italy

ABSTRACT

Ion irradiation has been used to transform spherical bimetallic AuAg nanocluster embedded in silica in a more complex structure made of a central cluster surrounded by a halo of smaller satellite nanoclusters, whose composition, size and distance from the central cluster can be tailored by controlling the irradiation parameters. This peculiar topology produces a red-shift of the surface plasma resonance of the composite through the electromagnetic coupling between the central cluster and the satellites. A calculation of the local field properties of the investigated systems within the fully-interacting generalized Mie theory showed that the satellite topology produces large local field enhancements around the central cluster.

INTRODUCTION

Monoelemental or metal alloy nanoclusters embedded in SiO_2-based matrices exhibit peculiar nonlinear optical properties which are function of the cluster size and composition [1-5]. Among different synthesis techniques, sequential ion implantation in glass has demonstrated to be a very effective technique to obtain such nanocomposites [6]. In this work, we used an ion beam-based multi-step approach for synthesizing embedded nanoclusters and for modifying in particular their near-field and far-field properties, through an ion-beam controlled tuning of the dielectric environment around them, therefore obtaining the plasmon tuning of the composite. In the first step noble metal nanoclusters are synthesized in silica by ion implantation followed by thermal treatments. The second step is ion irradiation which allows to create a peculiar nanostructure made of a halo of small satellite nanoclusters around the original ones [7,8]. For instance, in the Au-Cu system we found that irradiation with Ne^+ ions promotes a preferential extraction of Au from the alloy, resulting in the formation of Au-enriched "satellite" nanoparticles around the original Au_xCu_{1-x} clusters [7]. Despite the experimental demonstration of this core-satellite formation in ion irradiated mono- and bi-elemental systems, the microscopic mechanisms triggering the process are still not thoroughly understood. Therefore, we performed a systematic investigation of the role played by the irradiation parameters (i.e., fluence, flux, energy of the implanted ions) on controlling the satellite nanostructure.

In the present work, we used this ion beam based technique to promote a controlled red-shift of the plasma resonance absorption (i.e., plasmon tuning) of bimetallic Au_xAg_{1-x} clusters exploiting the coupling of the satellite nanoclusters with the original clusters. The obtained shift has been interpreted in the frame of fully-interacting generalized multiparticle Mie theory.

EXPERIMENTAL

Room temperature implantations were performed with a 200 keV high-current implanter (Danfysik 1090) at the INFN-INFM Implantation Laboratory (Legnaro, Italy). Fused silica (type II, Heraeus) slides were sequentially implanted with Au^+ and Ag^+ ions at a current density of 2 $\mu A/cm^2$, ion energies of 190 keV for Au and 130 keV for Ag. In order to produce large and well separated Au-Ag alloy clusters, the implanted slides were heat-treated in a conventional furnace in air at $800^{\circ}C$ for 1 h. This process produced $10 \div 20$ nm $Au_{0.6}Ag_{0.4}$ alloy nanoclusters buried at a depth of about 70 nm (the ion projected range) from the silica surface. This is the reference sample. The second step in the sample synthesis is a further ion irradiation to create the core-satellite cluster nanostructure. The effect of the irradiation parameters on the nanocomposites have been studied with three sets of irradiations in which the parameters varied were: (i) the irradiating ion (therefore changing the nuclear vs. electronic ratio of the stopping power). In this set, He^+, Ne^+, Ar^+ or Kr^{2+}, were used by tuning implantation energy and current densities to release to the sample the same power and energy density, averaged over the whole ion range. Then 190 keV Ar^+ ions were used for irradiation varying (ii) the fluence ($1 \div 5 \times 10^{16}$ cm^{-2}) and (iii) the current density ($0.2 \div 10$ $\mu A/cm^2$). Structural and compositional characterization was performed by transmission electron microscopy (TEM) at CNR-IMM Institute (Bologna, Italy) with a field-emission gun (FEG) microscope (FEI Tecnai F20 Super Twin) operating at 200 kV equipped with an EDAX energy-dispersive x-ray spectrometer (EDS) and a Gatan STEM controller for performing scanning transmission microscopy (STEM). Optical absorption spectra were collected with a dual beam spectrophotometer (CARY 5E UV–VIS–NIR) in the 200–800 nm wavelength range. Gold and silver concentration profiles were measured by 2.2 MeV $^4He^+$ Rutherford back-scattering spectrometry (RBS).

DISCUSSION

The conditions used in the present work for irradiating the bimetallic AuAg nanoclusters in silica were chosen to vary the nuclear vs. electronic component of the ions energy loss. This was obtained using different ions and irradiation energies to have projected ranges (R_P) of about 200-250 nm, i.e., about 3 times the projected range of the implanted metallic species (about 70 nm for both Au and Ag). This allows to safely rule out strong concentration overlap between the irradiating and implanted ions. In the following, we focused our analysis on the largest (about 25 nm in size) bimetallic alloy $Au_{0.6}Ag_{0.4}$ nanoclusters near R_P.

RBS analysis indicated that upon irradiation there is only a negligible rearrangement of the implanted species, indicating that the modification produced is purely local. Figure 1(a) shows the bright-field TEM cross-sectional image of the $Au_{0.6}Ag_{0.4}$ nanoclusters near R_P in the reference AuAg sample in comparison with that of the sample irradiated with 380 keV Kr^{2+}, at a fluence of 1.2×10^{16} ions/cm^2 (Fig. 1(b)). The most evident result of the multistep ion processing is the new morphology obtained: after ion irradiation we clearly observe the formation of 2-5 nm satellite clusters around each original cluster (similar images are obtained after irradiation with He, Ne and Ar ions) [9-11].

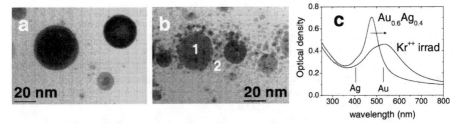

Figure 1. Bright-field TEM cross-sectional micrograph of the sample AuAg before (a) and after irradiation at room temperature with 380 keV Kr^{2+}, at a fluence of 1.2×10^{16} ions/cm^2 and current density of 0.4 μA/cm^2 (b). The corresponding optical absorption spectra of both samples are shown in (c): as a reference, vertical bars, labeled Ag and Au, indicate the SPR absorption peaks of pure Ag and Au nanoclusters in silica, respectively.

This change in morphology produces a dramatic change also of the linear optical absorption spectrum of the samples, in particular for the sample irradiated with 380 keV Kr^{2+} ions at a fluence of 1.2×10^{16} ions/cm^2, which have the largest nuclear component of the energy loss with respect to irradiation with the other lower mass ions. The results are shown in Fig. 1(c): the reference sample exhibits a surface plasma resonance (SPR) band centered at about 475 nm (due to the Au-Ag alloy formation), whereas after Kr^{2+} irradiation the band results red-shifted, doubly peaked (with one main resonance at 538 nm and a shoulder at 495 nm) and more damped with respect to the unirradiated sample.

Detailed TEM-EDS compositional analysis with a focused electron beam of the FEG-TEM on the irradiated samples ruled out the possibility to correlate the red-shift of the optical absorption band to a simple change in the nanoclusters composition. Indeed, with respect to the original atomic ratio of Au/Ag = 1.4 ± 0.2 of the nanoclusters in the unirradiated sample, the composition of the central clusters in the Kr-irradiated sample (zone 1 in Fig.1b) is not changed. On the other hand, the same ratio measured on the satellite clusters (zone 2 in Fig.1b) gave an Au/Ag atomic ratio of 4.0 ± 0.4. This preferential extraction of Au with respect to Ag is a general results found in all the investigated irradiations performed with He, Ne and Ar. Therefore, besides the creation of the new morphology (core-satellites clusters), we found a second and perhaps more intriguing feature of ion irradiation of metal nanoclusters embedded in silica, i.e. the elemental-selective extraction of Au and Ag from the alloy nanoclusters, similarly to what we obtained in Ne^+ irradiated Au-Cu nanoclusters [7].

A simple interpretation of the irradiation-induced modification of the optical absorption could be the convolution of two absorption contributions: the peak at shorter wavelength could be the residual of the original absorption peak of AuAg alloy nanoclusters, while the second could be ascribed to the Au-rich satellite clusters. We have ruled out this possibility considering the following arguments: (i) TEM-EDS compositional analysis on the satellite clusters indicates an $Au_{0.8}Ag_{0.2}$ average alloy composition in the case of Kr-irradiation. The SPR position in the Au_xAg_{1-x} alloy is a monotonically increasing function of the Au fraction from that of pure Ag (405 nm) to that of pure Au (530nm). Therefore, the satellite clusters should have a SPR peak position appreciably smaller than that of pure Au clusters. On the contrary, the measured position of the second band is about 540 nm, i.e., about 8-10 nm larger. (ii) TEM analysis indicates that the size and composition of the original clusters are slightly modified upon irradiation, because the number of atoms extracted from the original clusters amount to some percents of the total. This implies that, after irradiation, one should see the SPR of the core nanoclusters almost unchanged as compared to the original absorption spectrum, i.e., with a similar intensity ratio with respect to the interband transitions contribution (the absorption below 350 nm). On the contrary, while the interband absorption is unchanged after irradiation, the band near 490 nm is severely damped with respect to the original SPR peak.

Based on these two arguments, we concluded that a simple superposition of the absorption spectra of two non-interacting classes of nanoclusters can not account for the observed spectra. Therefore, to better understand the nature of the red-shift and of the doubly peaked absorption band, we have used a fully-interacting approach in which the simple Mie theory is replaced by an extension including explicitly the interaction between neighboring nanoclusters, i.e., the generalized multiparticle Mie (GMM) theory. Extensive calculations [12] show that it is possible to explain both features by considering the additive contribution of two polarization of the incident light: the red-shifted peak at 538 nm arises mainly from polarizations parallel to the plane of the nanoclusters at R_P which became coupled by the overlapping satellite halos of adjacent nanoclusters, promoted by the nuclear component of the energy loss under Kr irradiation. The shoulder at lower wavelength (495 nm) can be assigned, on the other hand, to the polarization state normal to the plane (i.e., along the surface normal), similarly to what happens in ellipsoidal particles in the Gans theory. It is interesting to note that the GMM approach is able to give insights also on the local field properties of the investigated systems. In particular, we found that the core-satellite cluster morphology produces local-field enhancements near the resonances as high as 15-20, with field hot-spots in the nanometric regions between the core cluster and the surrounding satellites.

By properly tuning the irradiation conditions it is therefore possible to increase the electromagnetic coupling within the produced structures. Considering that not only the linear but also the nonlinear properties of nanoclusters embedded in silica are highly sensitive to the local field properties, our multistep approach offers the possibility to accurately control the local field enhancement factors without modifying too much the original nanocluster size or composition.

In a previous work [10] we demonstrated that the number of atoms ejected from the $Au_{0.6}Ag_{0.4}$ clusters can be increased by increasing the nuclear fraction of the energy loss of the implanted ions. Indeed, the material emission from the $Au_{0.6}Ag_{0.4}$ cluster can be correlated to the formation in the original cluster of vacancies whose number increases with the nuclear fraction of the energy loss (as obtained by simulations with the SRIM package). This indicates that the formation of satellite nanoclusters can be understood in terms of ballistic processes induced by the collisional cascades. Moreover, the TEM analyses on the samples irradiated at increasing fluences indicated that the flux of ejected atoms from each original cluster surface is a linear function of the radiation fluence, as one can expect during a recoil mixing process [13].
The irradiating ions fluence plays also an important role to modify the size and the volumetric density ρ_{SC} of the satellite clusters. The satellite cluster size initially increases with fluence and then tends to saturate for larger fluences. At the same time a reduction of satellite cluster volumetric density ρ_{SC} around each original cluster is observed. This can be understood by assuming the onset of an Ostwald ripening process. The satellite clusters formation process could therefore be divided into three different stages: in the first one Au and Ag atoms are ejected from original alloy clusters following a ballistic process. Then Au and Ag atoms diffuse into the

matrix due to the mixing and to the radiation enhanced diffusion, starting the nucleation and growth where their concentration overcomes their solubility limit. In the last stage a different regime in the growth starts following an Ostwald ripening regime [14]. To experimentally support the interpretation of the vacancy formation as a central mechanism in the core-satellite formation, we add that inside some clusters irradiated at high current density the presence of voids was clearly visible in the TEM images, probably linked to clustering of the vacancies formed during the collisional cascades.

CONCLUSIONS

In the present work, we have used a multistep ion beam based technique to promote a controlled red-shift of the plasma resonance absorption of AuAg alloy nanoclusters (plasmon tuning) exploiting the coupling of the satellite nanoclusters with the original clusters. Due to the remarkable enhancement of the local fields (that we have simulated within a Generalized Multiparticle Mie approach), the obtained nanocomposite glass can be very promising for applications in nonlinear optics.

ACKNOWLEDGMENTS

The technical assistance of Dr. Carlo Scian during ion implantation processes is gratefully acknowledged. This work has been partially supported by Italian MIUR (University and Research Ministry) within a National University Research Project (PRIN 2004).

REFERENCES

1. P. Mazzoldi, G.W. Arnold, G. Battaglin, F. Gonella, and R.F. Haglund, *Nonlin. Opt. Phys. Mat.* **5**, 285(1996)
2. G. Mattei, *Nucl. Instr. and Meth. B* **191**, 323 (2002)
3. F. Gonella, G. Mattei, P. Mazzoldi, C. Sada, G. Battaglin, and E. Cattaruzza, *Appl. Phys. Lett.* **75**, 55 (1999)
4. R.H. Magruder III, J.E. Wittig, R.A. Zuhr, *J. Non-Cryst. Solids* **163**, 162 (1993)
5. E. Cattaruzza, G. Battaglin, F. Gonella, G. Mattei, P. Mazzoldi, R. Polloni, B.F. Scremin, *Applied Surface Science* **247**, 390 (2005)
6. P. Mazzoldi and G. Mattei, *La Rivista del Nuovo Cimento* **28**, 1 (2005)
7. G. Mattei, G. De Marchi, C. Maurizio, P. Mazzoldi, C. Sada, V. Bello, and G. Battaglin, *Phys. Rev. Lett.* **90**, 085502 (2003)
8. G. Rizza, H. Cheverry, T. Gacoin, A. Lamasson, S. Henry, *J. Appl. Phys.* **101**, 14321 (2007)
9. G. Mattei, V. Bello, G. Battaglin, G. De Marchi, C. Maurizio, P. Mazzoldi, M. Parolin, and C. Sada, *J. Non-Cryst.Solids* **322**, 17 (2003).
10. V. Bello, G. De Marchi, C. Maurizio, G. Mattei, P. Mazzoldi, and C. Sada, *J. Non-Cryst. Solids* **345-346**, 685 (2004).
11. G.Mattei, V. Bello, P. Mazzoldi, G. Pellegrini, C. Sada, C. Maurizio, G. Battaglin, *Nucl. Instrum. and Meth. B* **240**,128 (2005)
12. G. Pellegrini, V. Bello, G. Mattei, P. Mazzoldi, submitted.
13. J.C. Pivin and G. Rizza, *Thin Solid Films* **366**, 284(2000)
14. A. Miotello, G. De Marchi, G. Mattei, P. Mazzoldi, and C. Sada, *Phys. Rev. B* **63**, 075409 (2001)

Mater. Res. Soc. Symp. Proc. Vol. 1020 © 2007 Materials Research Society   1020-GG06-11

Pulsed Low-energy Ion-beam Induced Nucleation and Growth of Ge Nanocrystals on SiO₂

Anatoly Dvurechenskii[1], Nataly Stepina[1], Pavel Novikov[1], Vladislav Armbrister[1], Valery Kesler[1], Anton Gutakovskii[1], Victor Kirienko[1], Zhanna Smagina[1], and Reiner Groetzschel[2]
[1]Siberian Branch of Russian Academy of Sciences, Institute of Semiconductor Physics, Novosibirsk, 630090, Russian Federation
[2]Forschungszentrum Rossendorf, Dresden, D-01314, Germany

ABSTRACT

Pulsed low-energy (200 eV) ion-beam-induced nucleation during Ge deposition on thin SiO₂ film was used to form dense homogeneous arrays of Ge nanocrystals. The ion-beam action is shown to stimulate the nucleation of Ge nanocrystals when being applied after thin Ge layer deposition. Temperature and flux variation was used to optimize the nanocrystal size and array density required for memory device. Kinetic Monte Carlo simulation shows that ion impacts open an additional channel of atom displacement from a nanocrystal onto SiO₂ surface. This results both in decrease of the average nanocrystal size and in increase of nanocrystal density.

INTRODUCTION

The use of nanocrystals (NCs) embedded in a dielectric matrix has been widely studied for diverse applications for a good number of years. Non-volatile memories are one of the attractive areas of applicability for semiconductor NCs. An advantage of the nano-floating gate memory over the continuous floating gate is its improved endurance due to preventing lateral charge transport. Faster writing/erasing time, lower operating voltage and longer retention time have been demonstrated in memory device based on Si NCs embedded in SiO₂ [1,2]. Many research activities have been carried out on NC Si and Ge films grown on SiO₂ substrates using, for example, molecular beam epitaxy (MBE) [3], chemical vapor deposition (CVD) [4], and

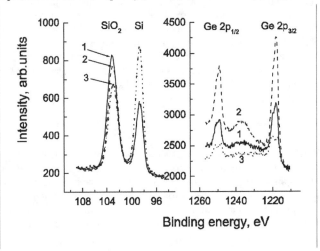

Figure 1. ESCA spectra for Ge deposition at 300°C with usual MBD (1) and ion-assisted MBD for "capped SiO₂" (2) and "uncapped SiO₂" (3) regimes. Left - peak of 2p Si, right - peak of 2p Ge.

pulsed-laser deposition [5]. Various methods were used for fabrication of Si and Ge NCs embedded in dielectric matrix, such as ion beam synthesis [6], oxidation and reduction of Ge/Si NCs [7], rapid thermal annealing of cosputtered layers [8]. Recently King showed that Ge has

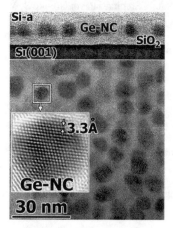

Figure 2. HREM image (upper part - cross-section, lower part - plan view) of nanocrystals on SiO_2 J_{Ge}=0.08 ML/s, T=250°C.

the superior properties over Si as a charging nodes in a single-electron memory device (SEMD) in terms of the writing/erasing time and the operating voltage [4]. However, space distribution of NCs within dielectric obtained by most of these techniques is random. To suppress an undesirable effects resulted from the tunneling distance fluctuation, one should form in-plane arrangement of NCs. Another requirements that can be beyond the control are the size of NCs, their density and homogeneity in growth plane. That is the challenge for most of above-mentioned growth methods. Our preliminary results [9] showed the stimulation effect of pulsed low-energy (100-200 eV) ion-beam action on nucleation and growth of Ge NCs on relatively thick (~100 nm) SiO_2 films during Molecular Beam Deposition (MBD).

To understand the mechanism of pulsed ion beam influence on nucleation process and sputtering of SiO_2 layer we studied different regimes of Ge NCs arrays growth on thin SiO_2 films typically used as tunnel oxide in NC-based memory devices.

EXPERIMENT

A 3.5 nm thick SiO_2 film was grown by thermal oxidation on (111) p-type silicon substrates (1 Ω·cm) at 850 °C. After dioxide formation the wafers were washed, dried-up and inserted into the ultrahigh-vacuum (UHV) chamber of molecular beam epitaxy setup equipped with effusion cell for Ge. The system of ionization and acceleration of Ge^+ ions provided the degree of ionization of Ge molecular beam from 0.1% to 0.5%. MBD was carried out at temperature varied from 250 to 400 °C. The total Ge effective thickness of deposition was 20 monolayers (ML). The rate of Ge deposition was varied between 0.08-0.19 ML/s. The generated ion-current pulses had duration of 0.5-1 s with ion energy of 200 eV. Two different regimes of

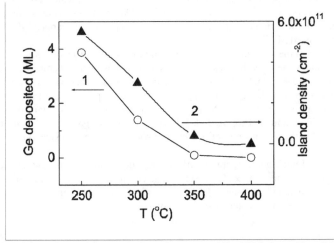

Figure 3. NCs density (HREM data - 2) and amount of Ge (RBS data - 1) as dependent on growth temperature. J_{Ge}=0.125 ML/s.

ion beam action were used. In the first case (called "capped SiO_2" regime) three Ge ML were deposited without ion-beam action. Pulsed ion-beam actions were applied in series at the effective Ge layer thickness of 3 ML, 4 ML and 5 ML. In the second case (called "uncapped SiO_2" regime) ion implantation started simultaneously with Ge deposition and repeated after deposition of 1 ML and 2 ML. To prevent oxidation of Ge some samples were covered by thin (~5 nm) a-Si layer before taking them out from UHV chamber.

The chemical content of the samples with open Ge NCs surface were studied by Electron Spectroscopy for Chemical Analysis (ESCA) after transferring the samples via ambient atmosphere to ESCA spectrometer. Fig.1 shows the detailed spectra of Ge 2p doublet and Si 2p for samples grown by usual MBD (curve 1) as well as by ion-beam induced nucleation at "capped SiO_2" (curve 2) and "uncapped SiO_2" (curve 3) surface at T=300 °C. The analysis of the intensity of Si 2p line shown in the left part of Fig.1 has allowed us to clarify the ion beam effect on tunnel oxide layer during Ge deposition. The thickness of tunnel oxide was evaluated from the intensity ratio of the oxide component Si 2p (103.0 eV) to the substrate component Si 2p (98.7 eV). The significant decrease in oxide thickness (from 4.2 to 2.4 nm) is observed in case, when ions irradiate an uncapped SiO_2 surface. Conventional and ion-assisted MBD at capped SiO_2 surface keep dioxide layer thickness practically unchanged. In all cases we found no changes in stoichiometry of SiO_2 layer. Amount of Ge remaining at SiO_2 surface can be determined from the spectra of Ge 2p doublet shown in the right part of Fig.1. Both Ge $2p_{1/2}$ and Ge $2p_{3/2}$ lines show pronounced superposition of two components with energy shift of 2.3 eV related to germanium in elemental and partially oxidized states, respectively. Precise analysis of Ge 2p and Ge 3d photoelectron lines characterized by significant difference in electron mean-free-path gave an evidence, that GeO_x caps the pure germanium nuclei. Obviously, the GeO_x fraction originates from the transferring the samples to ESCA spectrometer via ambient atmosphere. Concentration of Ge deposited was determined from the integral intensity of Ge $2p_{3/2}$ peak. Both usual and ion-assisted MBD at "uncapped SiO_2" show smaller amount of Ge remaining on the SiO_2 surface than that for "capped SiO_2" regime, which gives 1.7 times higher Ge content. However, even in this case only ~30% of Ge remains on SiO_2 surface, indicating strong Ge desorption during the growth. Later on, we keep in mind "capped SiO_2" regime when mention ion-assisted MBD.

Fig.2 shows the structure of Ge NCs for High Resolution Electron Microscopy (HREM) cross-section and plane view images of samples prepared using the chemical etching from the substrate side of the sample. Interplanar spacing corresponds to that for (111) planes in Ge.

Temperature and Ge flux were varied to optimize the parameters of NCs ensemble. Homogeneous array of Ge ball-like NCs with ~6 nm average size was obtained by MBD using irradiation at 250 °C and Ge flux J_{Ge} 0.08 ML/sec. Influence of growth temperature on NCs density (HREM data) and Ge amount (Rutherford backscattering (RBS)data) is presented in Fig.3.

We have found that temperature affects only the density of this ensemble keeping the island size practically unchanged. Ge desorption is enforced with the increase of the substrate temperature. The detailed analysis of HREM data shows, that ion beam nucleation suppresses Ge desorption, decreases NC size and size dispersion (Fig.4).

DISCUSSION

To clarify the mechanism of Ge NCs formation and the role of ion irradiation we have carried out the kinetic Monte Carlo (MC) simulation. The kinetics of NCs formation was calculated using "lattice gas" model [10], which includes Ge atoms deposition on SiO_2 surface, their surface diffusion, desorption, precipitation, and ion-beam action. The sites occupied by Ge atoms and SiO_2 were restricted to faced-centered cubic (FCC) lattice, which is well suited to

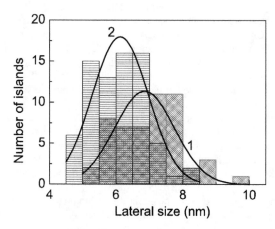

Figure 4. NCs size distribution for the irradiated (2) and non-irradiated (1) samples grown at T=250°C and Ge flux J_{Ge}=0.1 ML/s.

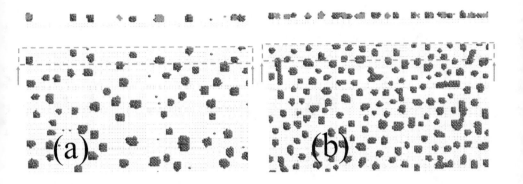

Figure 5. Ge NCs on SiO$_2$ simulated by MC without (a) and with (b) ion-beam action. The upper image is a cross-section of the same array within the region marked with a dashed rectangle on the plan view. T=350°C, J$_{Ge}$=0.1 ML/s. NCs size distribution for the irradiated (2) and non-irradiated (1) samples grown at T=250°C and Ge flux J$_{Ge}$=0.1 ML/s.

describe precipitation in isotropic amorphous matrices (e.g. SiO$_2$). The lattice includes 128×128×32 sites with cylindrical boundary conditions in the lateral plane. The number and types of nearest neighbors (NN) to a Ge atom determine its interaction energy with other Ge atoms and SiO$_2$ matrix. The energy per one Ge-Ge bond in FCC lattice was taken equal to 0.2 eV, which corresponds to an activation surface diffusion energy 0.8 eV at free Ge surface obtained from molecular dynamics calculation [11]. For Ge in SiO$_2$ environment the interaction energy per one neighbor site occupied by SiO$_2$ was chosen as much as 0.076 eV, which provides the reproduction of the experimental dependence of Ge amount remaining in the surface upon the Ge amount deposited. The kinetics of Ge atoms on the SiO$_2$ matrix surface is simulated by jumps to empty sites. The probability P of each jump is:

$$P \propto e^{-\frac{(n_i - n_f)E}{kT}}, \tag{1}$$

where n_i and n_f are the numbers of neighbors of the jumping atom in initial and final position, respectively; E is a bond energy per one neighbor; k is Boltzmann constant; T is the temperature.

Fig.5a presents the results of MC simulation of Ge NCs (plan-view and cross-section images) formed on SiO$_2$ without ion-beam action. The NCs have isotropic shape, corresponding to minimal surface energy, as would be expected for thermal equilibrium condition at nonwettable SiO$_2$ surface. The growth is a kinetic rather than equilibrium process. However, our calculations showed that the surface energy minimization remains a dominating factor of NCs formation. In the frame of this model an irradiation effect was described by the concept of collisional mixing (CM), i.e. the displacements of atoms. In the simulation process each Ge atom has a probability P_{DP} to be displaced by a distance R during one MC step:

$$P = qe^{-\frac{R}{\lambda}} , \qquad (2)$$

where λ is the length of displacement, depending on an ion energy, q is the factor proportional to the ion flux density. We neglected the known effects of temperature-dependent ion-solid interactions, which are observed at temperatures above 300 °C and ion energies of about 100 eV and lower [12]. Ion-beam action results in the smaller NCs size and the higher NCs density (Fig.5b) due to precipitation and nucleation of new NCs by atoms knocked out from initial NCs to the SiO_2 surface.

CONCLUSIONS

In conclusion we have shown that pulsed low-energy ion-beam-induced nucleation and growth during Ge deposition on SiO_2 films allow to suppress Ge desorption, increase the density of Ge nanocrystals, decrease the average nanocrystal size and size dispersion.

ACKNOWLEDGMENTS

This work was supported by the Russian Foundation for Basic Research (Grant No.06-02-08077).

REFERENCES

1. S. Tiwari, F. Rana, H. Hanati, A. Hartstein, E.F. Crabble and K. Chan, Appl. Phys. Lett. **68**, 13770 (1996).
2. D.W. Kim, T. Kim and S.K. Benerjee, IEEE Trans. Electron Devices **50**, 1823 (2003).
3. A. Kanjilal, J.L. Hansen, P. Gaiduk, A.N. Larsen, N. Cherkashin, A. Claverie, P. Normand, E. Kapelanakis, D. Skarlatos, and D. Tsoukalas, Appl. Phys. Lett. **82**, 1212 (2003).
4. Y.C. King, T.J. King, and C. Hu, IEEE Trans. Electron Devices **48**, 696 (2001).
5. X.B. Lu, P. F. Lee and J.Y. Dai, Appl. Phys. Lett. **86**, 203111 (2005).
6. P. Normand, E. Kapetanakis, D. Tsoukalas, G. Kamoulakos, K.Beltsios, J.Van Den Berg, S. Zhang, Mat. Sci. and Eng. C **15**, 145 (2001).
7. T. Sass, V. Zela, A. Gustafsson, I. Pietzonka and W. Seifert, Appl. Phys. Lett. **81**, 3455 (2002).
8. T. Sass, V. Zela, A. Gustafsson, I. Pietzonka and W. Seifert, Appl. Phys. Lett. **81**, 3455 (2002).
9. A.V. Dvurechenskii, P.L. Novikov, Y. Khang, Zh.V. Smagina, V.A. Armbrister, A.K. Gutakovskii, Proc. SPIE, 2006, **6260**, 626006-1 (2006).
10. P. Novikov, K.-H. Heinig, A. Larsen, A. Dvurechenskii, Nucl. Instum. Methods Phys. Res. B **191**, 462 (2002).
11. D. Srivastava and B.J. Garrison, Phys. Rev. **B46**, 1472 (1994).
12. Z. Wang and E.G. Seebauer, Phys. Rev. Lett. **95**, 015501 (2005).

Mater. Res. Soc. Symp. Proc. Vol. 1020 © 2007 Materials Research Society   1020-GG04-04

Time Evolution of Nano Dots Created on InP(111) Surfaces by keV Irradiation

Dipak Paramanik, Subrata Majumdar, Smruti Ranjan Sahoo, and Shikha Varma
XPS/SPM Laboratory, Institute of Physics, Sachivalay Marg, Bhubaneswar, India

ABSTRACT

Fabrication of Nanodots on semiconductor surfaces has immense importance due to their application in memory and optoelectronic devices. Ion irradiation methods display an easy and cost effective route for developing self assembled structures. We have studied the formation of Nano-dots on InP(111) surfaces by 3keV Ar ion irradiation. The distribution of nano Dots on InP surfaces has been investigated by Scanning Probe Microscopy (SPM). A 5 min irradiation of InP surface with Ar ions leads to the appearance of dots on the surface. The density of dots is, however very small. These dots have been obtained at room temperature, in the absence of sample rotation, with an angle of 15 degree between the ion axis and the sample normal. After an irradiation of 10 min a large density of dots appear on InP surface and display a narrow distribution of size and height. The dots at this stage have an average diameter of 25nm and a height of 4nm. With increased irradiation time the average size and the height of the dots increase and their distributions become broader. This scenario, however, changes after 40 min irradiation where large rectangular shaped dots of about 100 nm diameter and 40 nm height are observed. Surprisingly, for larger irradiation times a reduction in the size and height is observed. The studies suggest presence of "critical time" (t_c) at t= 40 min such that the dot structures grow with time below t_c but diminish in size beyond it.

INTRODUCTION

With the first growing interest in nanotechnology, fabrication of regular semiconductor nanostructure with controlled size and height is of great importance[1,2]. InP is a vastly used semiconductor material in this area. Ion beam sputtering is frequently regarded as an alternative process for the fabrication of various nano-structured surfaces or interfaces via self organization. Under certain conditions, sputtering can modify the surfaces resulting in pronounced topographic evolution producing well ordered pattern. This pattern formation is related to the surface instability caused by curvature dependent ion sputtering that roughens the surface and smoothing by different relaxation mechanism [3,4]. Recent studies show the formation of ordered InP nanostructure by Ar^+ ion sputtering under normal ion incidence [5,6] or alternatively under oblique ion incidence with simultaneous sample rotation[7]. Generally for off-normal ion incidence without sample rotation, a periodic height modulation in the form of ripple or wave like structure with a sub-micron length scale develops during ion bombardment as observed for single crystalline III-V semiconductors[8,9].

Starting from the Bradley-Harper (BH) theory [4] a successful description of the morphological evolution of the ion sputtered surface is described by the isotropic Kuramoto-Sivashinsky (KS) equation [5,6,10,11]. The temporal development of the surface profile h(x,y,t) is given by the following undamped KS equation:

$$\frac{\partial h}{\partial t} = -V_0 + \nu \nabla^2 h - D_{eff} \nabla^4 h + \frac{\lambda}{2} (\nabla h)^2 + \eta \,, \qquad (1)$$

where V_0 is the constant erosion velocity and v is the effective surface tension caused by the erosion process and usually has a negative value leading to a surface instability. D_{eff} is the surface diffusion coefficient which is the sum of thermal diffusion and ion induced effective diffusion. The nonlinear term $\lambda/2(\nabla h)^2$ accounts for the slope-dependent erosion yield and brings forth the saturation of surface roughness with time. The time where the surface roughness start to saturate is called the crossover time t_c. η is an uncorrelated white noise with zero mean, mimicking the randomness resulting from the stochastic nature of ion arrival to the surface.

In contrast, arrays of zero or two dimensional nanostructure can be produced by ion sputtering under normal incidence or alternatively by oblique incidence (>30° angle from sample normal) with simultaneous sample rotation[6,7]. Here we report for the first time the formation of order InP nanostructure for slight off-normal (15° angle from sample normal) ion incidence. SPM technique has been used to investigate the nano dots formation over wide temporal range and we observe the formation of most uniform dots with narrowest size distribution for 10 min sputtering. The average size and height of the nanostructures varies in the range of 10 to 100 nm and 5 to 40 nm, respectively, for different sputtering times. The nanostructures develope in the early stage of sputtering and grow in size and height with increasing time upto a certain critical time. Here, we find that for 40 min of sputtering duration the size and height of the nano dots become maximum. Thus, we find a crossover time t_c (40 min) up to which the size and the height of the nano dots increases with increasing sputtering duration. However, beyond t_c they start decreasing. At t_c rectangular and uniform celled arrays of nano dots are observed. For ion exposure durations larger than t_c an inverse ripening and fragmentation of bigger nano dots into smaller dots is observed. At this stage, surface diffusion may also become a dominant process. The rms surface roughness also increases upto t_c and decreases beyond it.

EXPERIMENT

Samples used in this work are commercially available epipolished InP(111) single crystal wafers. Sputtering were preformed in an ultra high vacuum chamber with a base pressure of 1×10^{-9} torr. InP wafers were placed on the sample holder with the help of double sided carbon tape and irradiated with Ar^+ ion beam from an EX05 ion gun from VG Microtech. The beam of 3 keV energy impinged on the surface at an incidence angle of 15° with respect to the sample normal. The beam was focused to a circular spot of 0.3 cm diameter on the sample surface with ion current density of 10uA/cm^2. Samples were sputtered for 5, 10, 20, 30, 40, 60 and 80 minutes. The fluence for one minute of sputtering is calculated to be 1.3×10^{16} ions/cm^2. The bombarded samples were studied in tapping mode in a Veeco Nanoscope IIIA multimode Scanning Probe Microscope (SPM). Several images of scan length 50 nm to 10 μm were acquired and analyzed.

RESULTS AND DISCUSSIONS

Fig 1 shows the two dimensional 500×500nm^2 SPM images of the surface that are sputtered for different times from 10 minutes to 80 minutes. For comparison, the virgin(unsputterd) sample is also shown here. Figure 2 shows the size and height distributions of the nanostructures for different sputtering time from 10 minutes to 80 minutes. In the early stages of sputtering i.e.

sputtering for 5 min the surface starts to become rough and mound or cone like structures begins to appear. The lateral size and height of these structures increase with large sputtering duration. For 5 min of sputtering the surface is dominated by small, wavy perturbation generated by the interplay between the ion induced instability and surface relaxation [5].

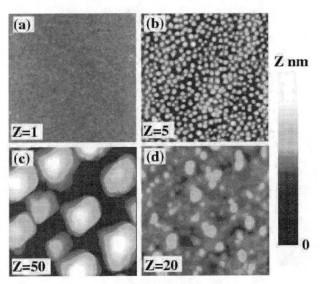

Figure 1. 500 nm × 500 nm SPM images of InP(111) surfaces for the (a) virgin (unsputterd) sample as well after 3 keV Ar$^+$ ions sputtering for duration (b) 10 min, (c) 40 min, and (d) 80 minutes. The color height scale bar with different Z values is also shown in the picture.

After sputtering for 10 minutes, these structures fully developed and turn into isolated nano dots(fig.1b). A highly uniform and narrow distribution of dots with an average size 25 nm and height 4 nm is observed . The density of dots is found to be 2×10^{11} cm^{-2}. For 20 minutes of sputtering, although the density of nanostructure does not change but the average size slightly increases and some overlapping among the nanostructures are seen. The average size and height of the nanostructures is 32 nm and 8 nm, respectively (fig.2). After 30 minute sputtering several small nanostructures agglomerate to form big structures. The average size and height of these nanostructure is 50 nm and height is 18 nm. Since several nanostructures join and form big size nanostructures the density of nanostructure decreases to 4×10^{10} cm^{-2}. The processes of ripening and agglomeration of nanostructures continues up to 40 minutes of sputtering. At 40 minutes uniformly distributed and regular shaped nanostructure are observed (fig.1c). Here most of the nanostructure are of rectangular shape with uniform orientation. Figure 2d show immense increase in the size and the height of the dots at this stage with their average size and height becoming 100nm and 45nm, respectively. The density of the dots is observed to be 6×10^{9} cm^{-2}. Interestingly, for sputtering durations higher than 40 min, a combination of inverse ripening and fragmentation of nano dots is observed. Fig.1d displays the fragmentation of several bigger dots into smaller dots after 80 min sputtering. As a consequence, the density of dots increases to

5×10^{10} cm^{-2} and the size of the dots reduces. The dots have an average size of 40 nm compared to 100 nm seen after 40min of sputtering. The average height of the dots is only 14 nm compared to 45nm after 40min sputtering. Moreover, small dots of sizes 10 nm and height 6 nm are also seen at this stage (Fig.2f), which were not seen after 40min. The decrease in height suggests that the inverse ripening may not just happen due to the fragmentation of dots but some mass flow from larger to smaller dots may also be involved. Both size and the height distributions are very broad at this stage. Existence of some large 100nm nano dots suggests that a few dots may also ripen at this stage. However, their number is very small. Fragmentation and inverse ripening, thus, dominate the structure formation at this stage.

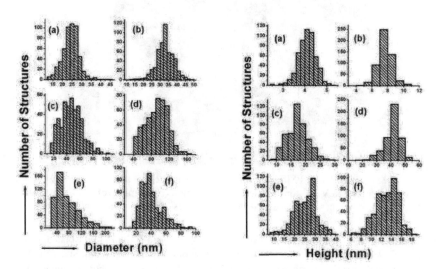

Figure 2. Size(diameter) and height distributions of nano dots formed after sputtering durations of (a) 10 min, (b) 20 min, (c) 30 min, (d) 40 min, (e) 60 min , and (f) 80 minutes.

Above study of nanostructure formation and distribution of their size and height for different sputtering time reveals that there is a crossover time t_c (40 min) where the regular and uniform nanostructures are formed. However for sputtering beyond t_c, the size and height of the nanostructures decrease and destroy the uniformity and regularity of their distributions get destroyed (fig.1 and 2). Using the KS model, *Kahng et al.* [5] theoretically showed that there is a crossover sputtering time t_c, where most uniform nanostructure are formed. For sputtering up to t_c the size and height of the nanostructure increase and become maximum at t_c. Although it was predicted theoretically a few years back this is the first experimental observation of crossover time during the sputtering process. The crossover behavior from linear to nonlinear regimes is also observed through the rms surface roughness. Fig.3 shows rms roughness of the InP surface for sputtering from 5 minutes to 80 minutes. This quantity exhibits a sharp transition at 40 minutes, which is the crossover time in the present study. For $t<t_c$, rms roughness increases from

0.5 nm for the virgin sample to 5.5 nm after a duration of 40 min sputtering, while for t>t_c it decreases and is about 2.9 nm after 80 min of sputtering. This smoothening behavior of the surface, for large sputtering durations, is surprising. Several studies[5,7] on InP surfaces demonstrate a saturation of surface roughness for long sputtering durations. However, none of these studies displayed any surface smoothening behavior. Interestingly in an earlier study, surface smoothening has also been observed by us for MeV irradiated InP surfaces [12] beyond the stage where crystal to amorphous transition occurs in the InP lattice [13].

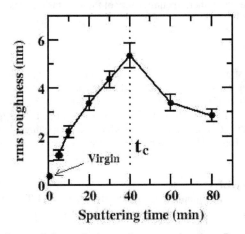

Figure 3. Variation of rms roughness for increasing sputtering time. Roughness for the virgin sample is also shown here. The critical time is marked at 40 minutes.

The decrease in surface roughness for longer time sputtering may be due to the fragmentation of bigger dots into smaller dots and due to the dominant diffusion process. Irregular patterns, characterized by the presence of mound or cone like structures, are induced on InP surface during the first 5 minutes of sputtering process. The surface is dominated by small, wavy perturbations generated by the interplay between the ion induced instability and surface relaxation [5]. The amplification of the random amplitudes by the negative surface tension competes with the smoothening processes such as surface diffusion and viscous flow. This leads to the formation of regular surface patterns after 10min of sputtering. The scenario that emerges shows that with the nucleation process starting at 5 min sputtering, regular and uniform patterns of nano dots appear after 10 min duration. During early stages of sputtering, upto t_c=40min, the dots ripen and agglomerate. This evolution is also accompanied by surface roughening. The surface at t_c is characterized by the presence of rectangular celled array of nano dots. Beyond t_c however, an inverse ripening and fragmentation of the dots take place in conjunction with the smoothening of surface. Diffusive smoothening may thus be controlling the surface patterning for sputtering durations larger than t_c. *Kahng et al.* had theoretically shown the existence of a critical time where most uniform and largest dots may be expected [5]. In the present study we observe, for the first time, the presence of the critical time characterized by ripening-agglomeration of dots for lower (t<t_c) and inverse ripening-fragmentation for larger(t>t_c) durations. Our results are however different than those seen in [5] where nano dots develop only at t_c . In the present study the most uniform nano dots with narrowest size distributions are formed far below t_c in the very early stages i.e. after 10min of sputtering. At t_c the dots are fully developed and are characterized by maximum size and height. At short times (t<t_c) the instability due to local surface curvature dependence of the sputter yield produces a regular array of dots and a rapid increase in roughness. At longer durations, beyond t_c, the non-linear effects are seen to drastically lower the growth of roughness. In the present study the non-linear effects cause the inverse ripening-fragmentation of the nanodots and lead to smoothening of the surface. Although the surface becomes smoother, the uniformity of the structures is destroyed as their distributions become broad.

CONCLUSIONS

We have investigated the formation of nano dots on InP(111) surfaces after sputtering with 3 keV Ar^+ ions in non- normal geometry and in the absence of rotation. The most uniform and narrowest distribution of nano dots is observed after 10min of sputtering. The study shows the existence of a critical time, t_c. The dots have been seen to ripen and agglomerate for sputtering durations smaller than t_c. For longer durations an inverse ripening, phenomenon not seen earlier, has been observed. Beyond t_c, ion induced enhanced diffusion causes the smoothening of the surface.

ACKNOWLEDGMENTS
This work is partly supported by ONR grant no. N00014-97-1-0991. We would like to acknowledge the help of S.K. Choudhury during sputtering.

REFERENCES

1. C.G. Smith, Rep. Prog. Phys. **59**, 235 (1996).
2. S. Facsko, T. Dekorsy, C. Koerdt, C. Trappe, H. Kurz, A. Vogt, H. L. Hartnagel, Science **285**, 1551 (1999)
3. Peter Sigmund, Phys. Rev. **184**, 383 (1969).
4. R. Mark Bradley and James M. E. Harper, J. Vac. Sci. Technol. A **6**, 2390 (1988).
5. B. Kahng, H. Jeong and A.-L. Barabasi, Appl. Phys. Lett. **78**, 805 (2001).
6. S.K. Tan, A.T.S. Wee, J. Vac. Sci. Technol. B **24**, 1444 (2006).
7. F. Frost, A. Schindler, and F. Bigl, Phys. Rev. Lett. **85**, 4116 (2000).
8. S.W. Maclaren, J.E. Baker, N.L. Finnegan, C.M. Loxton, J. Vac. Sci. Technol. A **10**, 468 (1992).
9. F. Frost, B. Ziberi, T. Hoche, B. Rauschenbach, Nucl. Instr. and Meth. B **216**, 9 (2004)
10. R. Cuerno, H.A. Makes, S.Tomassone, S.T. Harington, and A.-L. Barabasi, Phys. Rev. Lett. **75**, 4464 (1995).
11. M.Rost and J. Krug, Phys. Rev. Lett. **75** 3894 (1995).
12. Dipak Paramanik, Asima Pradhan, Shikha Varma, J. App. Phys., **99,** 014304 (2006).
13. Dipak Paramanik, Shikha Varma, J. App. Phys., **101**, 023528 (2007)

Examples: Applications and Devices

Mater. Res. Soc. Symp. Proc. Vol. 1020 © 2007 Materials Research Society   1020-GG07-02

Ion Bombardment Improvement on Thermoelectric Properties of Multilayered Bi2Te3/Sb2Te3 Deposited by Magnetron Sputtering

Bangke Zheng, S. Budak, B. Chhay, R. L. Zimmerman, and D. ILA
Department of Physics, Center for Irradiation of Materials (CIM), Alabama A&M University, P.O.Box 1447, Normal, AL, 35762

Abstract

We have grown the multi layer of Bi_2Te_3 and Sb_2Te_3 super-lattice film systems using the magnetron sputtering system. The purpose of using magnetron sputtering system is to o keep the stoichiometry of Bi_2Te_3 and Sb_2Te_3 so as to keep the electrical and thermal conductivity advantage of the layered structure of bulk Bi_2Te_3 and Sb_2Te_3. Magnetron sputtering is operated at relatively low temperature. The prepared super-lattice thin film systems were then bombarded by 5 MeV Si ions to form nano-clusters in the layers to increase the electrical conductivity and to decrease the thermal conductivity. We measured the cross plane electrical and thermal conductivities before and after MeV Si ions bombardment.

1. Introduction

As shown in Fig. 1, these thin films form a periodic quantum well structure consisting of tens to hundreds of alternating layers with different band gap. The thickness of each layer is 10 nm. The performance of super-lattice thin film thermoelectric device is quantified by the dimensionless figure of merit $ZT = S^2 \sigma T/ k$ [1]. Our aim is to obtain high ZT values by increasing the Seebeck coefficient S and the electrical conductivity σ, and reducing the thermal conductivity k by bombarding the super-lattice structure with MeV Si ions. The bombardment will form nanoscale cluster (quantum dot) structures.

In addition to the quantum well confinement of the phonon transmission due to Bragg scattering and reflection at lattice interface [1, 2, 3], the defect and disorder in the lattice caused by bombardment and the grain boundaries of these nano-scale cluster formed by bombardment increase the scattering of phonon and increase the chance of the inelastic interaction of phonon and the annihilation of phonon, inhibiting heat transport in the direction perpendicular to the lattice [4, 5, 6, 7]. Phonon is chiefly absorbed and dissipated along the lattice, thus cross plane thermal conductivity will decrease. These quantum dot clusters also increase the Seebeck coefficient and electrical conductivity due to the increase of the electronic density of states in nano-scale cluster miniband.

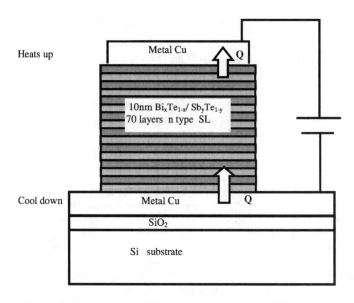

Figure 1. Schematic of $Bi_xTe_{1-x}/ Sb_yTe_{1-y}$ superlattice TE device

2. Experiment

The magnetron sputtering deposition system with two target holders (guns) was used to deposit Bi_2Te_3/ Sb_2Te_3 multilayer thin films. The chamber was pumped down by a turbo pump to a background pressure of 2×10^{-5} Torr. The two guns in our magnetron sputtering device are oriented at a certain angle to get off-axis plasma plume, which will form the lattice with preferential orientation for electric conductivity in each layer. The multilayer films were sequentially deposited on Si substrate that was coated with a SiO_2 insulation layer and a metal (Cu) contact layer to form a periodic nano-layer structure consisting of 70 alternating layers of $Bi_xTe_{1-x}/ Sb_yTe_{1-y}$. In this quantum-well super-lattice structure (SL), the thickness of each layer is 10 nm. After the deposition of these SL films, they were bombarded by 5 MeV Si ions using the AAMU Pelletron ion beam accelerator in order to form nanoscale cluster (quantum dot) structures. The SRIM simulation software shows that the Si ions with the energy value of 5 MeV go through $Bi_xTe_{1-x}/ Sb_yTe_{1-y}$ multilayer super-lattice films and terminate deeply in the substrate. Finally, a second Cu contact layer was deposited over the SL after bombardment. The samples were trimmed with a diamond cutter to obtain a clean edge of each layer. In this way, a complete thermoelectric (TE) device was made. The thickness of the deposited layer was controlled by a crystal oscillator thickness monitor. The parameter of DC magnetron sputtering deposition for Bi_2Te_3 is: Argon pressure is 1.2×10^{-2} Torr. For Sb_2Te_3 Argon pressure is 1.8×10^{-2} Torr.

Rutherford backscattering spectrometry (RBS) was used to determine the stoichiometry of Bi_xTe_{1-x} and Sb_yTe_{1-y} layer grown in this condition. Figure 2 shows the RBS spectrum and RUMP (RBS Analysis and Simulation Package) simulation results of a single Bi_xTe_{1-x} layer and

Sb$_y$Te$_{1-y}$ layer grown on Si substrate. The RUMP simulation result indicates that the Bi-Te layer and Sb-Te layer grown at this condition is characterized to be Bi$_2$Te$_3$, and Sb$_2$Te$_3$.

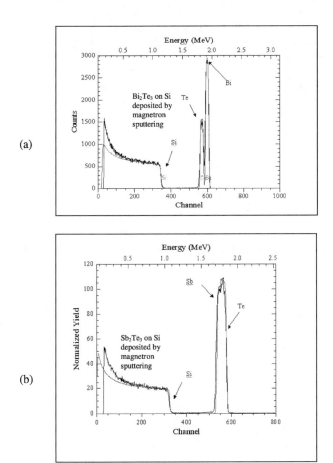

Figure 2. Rutherford Backscattering Spectrometry (RBS) and RUMP simulation of (a) single layer Bi-Te thin film and (b) single layer Sb-Te thin film on a Si substrate.

It is also possible to carry out low angle x-ray diffraction to determine the layer thicknesses in the multilayer, and determine the quality of the interface. After bombardment, this could be repeated to determine how the quality of the multilayer changes. We do not have a x-ray facilities but for the future ww think to find some colleagues to help us on this area to verify our RBS and RUMP results.

3. Thermoelectric Measurements and Results

We used the 3ω technique to measure the cross plane thermal conductivity of the thin film samples. The experimental setup has been previously described in refs.[3, 4, 8]. A narrow Pt strip is deposited onto the films providing a heater with a resistance value of about 200 Ω. Thermal conductivities of 70- layer Bi_2Te_3 / Sb_2Te_3 SL samples were measured before and after bombardment by 5 MeV Si ions with fluences of 0.5×10^{13}/cm^2, 1×10^{13}/cm^2, 5×10^{13}/cm^2, 1×10^{14}/cm^2 and these results are shown in figure 3.

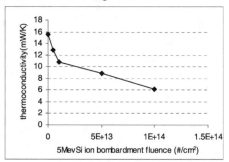

Figure 3. Thermal conductivity of 70 layers Bi_2Te_3/ Sb_2Te_3 before and after 5 MeV Si ions bombardment.

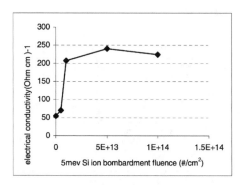

Figure.4 Cross plane electric conductivity of 70 layers Bi_2Te_3/ Sb_2Te_3 SL before and after 5 MeV Si ions.

We used a digital electronic bridge to measure the cross plane electrical conductivity of the thin film samples. We assumed that the Schottky junction barrier between Cu and the semiconductors is negligible and that the resistance of the copper surface electrodes is negligible. The electric conductivities of the 70 layers Bi_2Te_3/ Sb_2Te_3 samples were measure]\

d before and after 5 MeV Si ions bombardment and the results are shown in figure 4.

4. Discussion and Conclusion

Quantum well confinement effect is greatly enhanced by ion bombardment that formed nanocluster quantum dots. The defect and disorder in the lattice caused by bombardment and the grain boundary of these nano-scale cluster increase the scattering of phonon and increase the chance of the inelastic interaction of phonon and the annihilation of phonon, this limits phonon mean free path. This nanoscale cluster structure enhances phonon horizontal dissipation and absorption along the super-lattice rather than perpendicular to super-lattice, therefore decreases cross plane thermal conductivity. Although the disorder in the crystal resulted from bombardment in certain extent will decrease the electric conductivity, this tendency is overwhelmed by the increase of electronic density of state in the miniband of nanoscale cluster quantum dot structure formed by bombardment. After bombardment, the super-lattice crystal behaves more like an electronic crystal and phonon glass in the direction perpendicular to the superlattice interface. However, excessive bombardment will damage the structure of superlattice. The stoichiometry of Bi_2Te_3 and Sb_2Te_3 can be kept using magnetron sputtering deposition, which is operated at relatively low temperature, so that the electric and thermal conductivity advantage of the layered structure of bulk Bi_2Te_3 and Sb_2Te_3 is kept in each period of the super lattice

Acknowledgement

This research was sponsored by the Center for Irradiation of Materials, Alabama A&M University and by the AAMURI Center for Advanced Propulsion Materials under the contract number NNM06AA12A from NASA, and by National Science Foundation under Grant No. EPS-0447675.

References

[1] Xiaofeng Fan, Electronics Letters, 37 (2001) 126
[2] Xiaofeng Fan, Gehong Zeng, Appl. Phys. Lett., 78 (2001) 1580
[3] S.M Lee, D.G. Cahill, Appl. Phys. Lett., 70 (1997) 2957
[4] D.G. Cahill, M. Katiyar, Phys. Rev., B50 (1994) 6077
[5] Rama Venkatasubramanian, Phys. Rev., B61 (2000) 3091
[6] J. L. Liu, Phys Rev., B 67 (2003) 4781
[7] David G. Cahill, Rev. Sci. Instrum., 61 (1990) 802
[8] B. Zheng, S. Budak, C. Muntele, Z. Xiao, S. Celaschi, I. Muntele, B. Chhay, R.L. Zimmerman, L.R. Holland, and D. Ila, Mater. Res. Soc. Symp. Proc. Vol. 929 © 2006 Materials Research Society 0929-II04-12.

Mater. Res. Soc. Symp. Proc. Vol. 1020 © 2007 Materials Research Society   1020-GG07-05

Nanoscale Surface Modification of UltraHigh Molecular Weight Polyethylene (UHMWPE) Samples with the W + C Ion Implantation

E. Sokullu Urkac[1], A. Oztarhan[1], F. Tihminlioglu[2], N. Kaya[1], S. Budak[3], B. Chhay[3], C. Muntele[3], E. Oks[4], A. Nikolaev[4], and D. ILA[3]

[1]Department of Bioengineering, Ege University, Bornova, 35100, Turkey

[2]Department of Chemical Engineering, IYTE, Urla, 35000, Turkey

[3]Center for Irradiation of Materials, Alabama A&M University, 4900 Meridian Street, PO Box 1447, Huntsville, AL, 35762

[4]High Current Electronics Institute, Tomsk, 44000, Russian Federation

Abstract

In this work, Ultra High Molecular Weight Poly Ethylene (UHMWPE) samples were implanted by W + C ions using Metal-Vapour Vacuum Arc (MEVVA) ion implantation system with a fluence of 10^{17} ion/cm^2 and extraction voltage of 30 kV. Samples were characterized with Raman Spectra, ATR-FTIR, UV-VIS-NIR Spectrum and RBS . Surface morphology of implanted and unimplanted samples were examined in nanoscale with AFM.

Keywords: UHMWPE, W, C, Ion Implantation

1. Introduction

Materials used for orthopedic devices should have good biocompatibility, adequate mechanical properties, and sufficient wear and corrosion resistance, and they should be manufacturable at a reasonable cost. Most of total hip joints are composed of ultra-high molecular weight polyethylene (UHMWPE) because of this material's good biocompatibility, high resistance to wear and durability but wear debris is still the biggest problem thus, some of surface modifications are needed [1].

In this work we discuss a possible surface modification method, the ion implantation, performed by using ions of W+C to induce graphite-like surface structures in UHMWPE. The carbon–tungsten mixed-material system is probably the most widely studied system due to the formation of chemically and mechanically functionalized structures both experimentally and computationally [2]. In order to examine the effects of the implantation on UHMWPE surface, we did W+C hybrid ion implantation. Results of chemical characterization were presented and discussed before and after the treatment.

2. Material and Method

Samples with medical grade GUR 1020 - Type 1 - Ultra High Molecular Weight Polyethylene (UHMWPE - , Hipokrat Co. –CH$_2$–|$_n$ monomer) with a density of 945 kg/m^3 were used. Disk samples with a diameter of 30mm and thickness of 4mm were polished down to about surface roughness of 121,94(nm) Rms. Samples were hybrid implanted in gas and vacuum discharges with W+C ions by using MEVVA ion implanter with a fluence of 10^{17} ion/cm^2 and extraction voltage of 30 kV.

3. Characterization

In order to determine the ion penetration depth and ion ranges in the UHMWPE targets, TRIM simulation sotware was used. RBS analysis has been done with 2.1 MeV He ions incident at 30° and detection angle at 170°[3] .

In order to determine the damage products and structural changes in UHMWPE, additional analytical techniques were employed on the implanted samples. The samples were analyzed with optical absorption photospectrometry (OAP) for the ultraviolet–visible region. Optical characterization of samples before and after each treatment was carried out with a dual-beam spectrophotometer (Spectral Instruments). Attenuated total reflectance Fourier transform infrared (ATR-FTIR) measurements were performed using a Tensor 27 spectrometer with a Pike-ATR accessory equipped with ZnSe crystal. Raman spectra were acquired using a LabRam spectrophotometer using a He–Ne laser with $\lambda = 632$ nm which was focused on the sample surface by an Olympus BX-40 microscope using a 100× objective lens. The scattered light was collected in backscattering geometry by the same microscope. The results were compared with the identical untreated UHMWPE samples.

Atomic force microscopy (AFM) was used to investigate the surface morphology of the untreated and W+C implanted UHMWPE. A Digital Instrument- Multi-Mode-SPM apparatus was used to determine the surface roughness of untreated and Ag implanted UHMWPE samples. Scan size of 5.000μm, scan rate of 1.001 Hz and data scale of 300.0nm.were applied through silicon nitride tip for scanning the sample surfaces.

4. Results and Discussions

Range calculations as performed by TRIM yielded a rather symmetrical distribution as expected with the ionization of 14.5 % of W and 52 % of C, thus the distribution was truncated at the surface at a small fraction of the peak concentration. As seen in Figure 1 , the spectrum shows an O and Cu peaks after W+ C implantation. O is probably because of the oxidation reaction and the Cu peak is probably due to the contamination of the vacuum gas. The oxidation behaviors of UHWMPE after irradiation were reported in refs.[4,5]. The profitable influence of surface contamination from diffusion pump oil on tribological properties was also reported in ref. [6]. Cu peak that we have in RBS spectrum might be due to the the vacuum environment (pump oil or vacuum gas). Fig.2. shows the optical absorption photospectrum. As seen from fig. 2 W+C ion implantation leads to modification of its luminescence. The light emission in the near UV region practically disappears and a wide

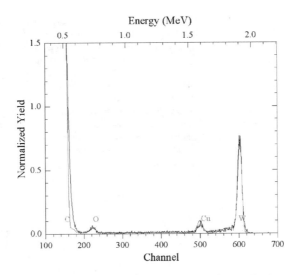

Figure 1. RUMP simulation of W+C implanted UHMWPE

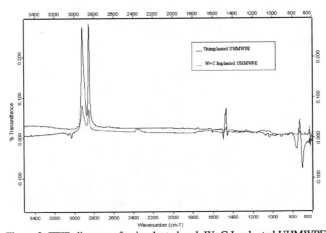

Figure 2. FTIR diagram of unimplanted and W+C Implanted UHMWPE

luminescence band develops at 500 nm. Rizatti et al. has observed the same behaviour with the ion bombarded poly (paraphenylene sulphide) films and they attributed these effects associated with a decrease in the optical gap (E_g) which can be related to a structural rearrangement of aromatic rings, possibly, generating a continuous network of conjugated cyclic structures [5].

In ATR-FTIR spectrum, the characteristic absorption bands for the CH_2 bonds appear in the 2900–2840, 1460–1370 and 740–720 cm^{-1} regions [6,7]. The transmission ATR analysis of the untreated and W+C implanted samples confirm the C–H band breaking since the C–H stretching (at 2847 cm^{-1}) band bending peaks of the pure UHMWPE sample disappears after W+C implantation. As seen in fig.3 the increase in the absorption bands in the 1594 and 1738,4 cm^{-1} regions, due to the implantation, has been attributed to the forming of unsaturated C=C bonds and a beginning of polymer oxidation [8,9]. One of the main products of the radiation-induced chemical changes is the C=C formation, and its stretching vibration is clearly observed in the spectra by an absorbance peak at 1594-1630 cm^{-1}. The presence of this peak suggests that after W+C implantation, the polymer surface becomes poor of hydrogen and rich of cross-linked carbon atoms. Furthermore, some carbonyl formation due to oxidative degradation was observed at around 1720 cm^{-1}.

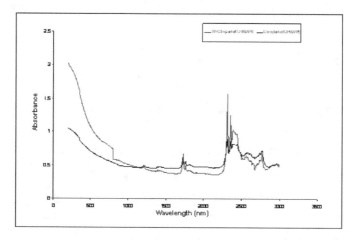

Figure 3. UV-VIS NIR Spectrum of Unimplanted and W+C Implanted UHMWPE.

The trans-CH wagging absorption band appears around at 965 cm^{-1} on the spectra of W+C implanted UHMWPE. Cis-CH wagging located at 610-66 cm^{-1} are also present for W+C implanted sample. Particulary, C-H characteristic peaks decrease with the implantation. In both the wagging vibrations, the motions of the hydrogens are partially balanced by countermotions of the substituents. This indicates that cis- & trans- geometric isomerism occured [10,11]. The formation of the peak at 3500 cm^{-1} region signifies the formation of NH and OH bonds at the higher end [10,11].

Raman spectrum of the pure (non–irradiated) UHMWPE sample exhibits the polyethylene's sharp characteristic peaks at 1060, 1127, 1293, 1440, 2722, 2846 and 2882 cm^{-1} [6]. From the baseline corrected Raman spectra, it is observed that the relative intensities of the bands at characteristic peaks all decreased as a result of the treatment. This fact suggests that the chemical structure of UHMWPE has changed after implantation[13]. Especially, the decrease of relative intensity of the peak at 1293 cm^{-1} indicates that the polymer chain is broken [14].

The two regions of interests in the Raman microprobe analysis are at 1595 cm^{-1} (G-line) and at 1350 cm^{-1} (D-line) are shown in Fig.4. The G-line is attributed to graphite

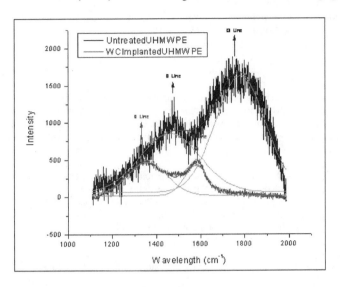

Figure 4. Raman Spectra of untreated and W+C implanted UHMWPE.

structure formation in the material and the D-line is attributed to amorphous structure or "disorder" in the material [6]. From the fitting, the position and full width at half maximum (FWHM) of the G and D peaks as well as the ratio of the D to G peak area—$I(D)/I(G)$, were determined. There is significant change in D/G ratio which is 0.612 for untreated sample and 1.88 for W+C implanted one. These observations indicate the formation of DLC layer on the UHMWPE surface as a result of ion implantation process. Fig. 5 shows the surface topography of W+C implanted UHMWPE sample. As seen from the Table 1, the surface roughness of the UHMWPE surface decreases from 121.94(nm) Rms before implantation and 41.28(nm) Rms after implantation. This shows that W+C implantation improves surface smoothness. It was reported that the rough surfaces caused more wear debris, therefore it is expected to have less wear debris after implantation of UHMWPE surfaces [15,16].

Figure 5. AFM image of W+C Implanted UHMWPE.

Table 1. Roughness Data of untreated and W+C Implanted UHMWPE obtained from AFM measurements.

Surface Roughness	Untreated UHMWPE	W+C Implanted UHMWPE
Rms (nm)	121.94	41,28

5. Conclusion

W+C hybrid ion implantation has been successfully applied for surface modification of medical grade UHMWPE. RBS spectra showed some O and Cu contamination formed at the surface. As a result of implantation the hydrogen content in the UHMWPE surface decreases, the chain structure is damaged and the modified surface layer is converted to hydrogenated amorphous carbon.

From the FTIR, Raman, RBS, and Optical Absorption Photosectroscopy results, we can infer that a new material was formed after bombardment. The new material has a structure cross-linked and constituted of conjugated bonds. UV absorption of W+C implanted UHMWPE increasing. These changes can be attributed to the irradiation effect of energetic ions. The AFM results show that surface roughness of implanted samples is decreased.

Acknowledgement
Authors acknowledge support from the Center for Irradiation of Materials, Alabama A&M University.

References

1. J. Black, G. Hastings, Handbook of Biomaterial Properties, 1998, Chapman & Hall.
2. R.P. Doerner, Journal of Nuclear Materials, 363-365, 32-40, 2007.
3. L.R. Doolittle, "Algorithms for the rapid simulation of Rutherford backscattering spectra", Nucl. Instr. and Meth. B, 9,344, 1985.
4. A. Turos, A.M. Abdul-Kader, D. Grambole, J. Jagielski, A. Piątkowska, N.K. Madi and M. Al-Maadeed, "The effects of ion bombardment of ultra-high molecular weight polyethylene", Nucl. Instr. and Meth. B, 249, 660-664,2006.
5. M.R. Rizzatti, M.A. Araújo and R.P. Livi, "The fluence effect of Ar^{++} bombardment in PPS", Nucl. Instr. and Meth. B, 174, 475-481,2001.
6. M. Braun,"Ion bombardment as a tool to modify surface properties of different materials", Vacuum, 38, 973-977,1988.
7. A.L. Evelyn, D. Ila , R.L. Zimmerman, K. Bhat, D.B. Poker and D.K. Hensley, "Effects of MeV ions on PE and PVDC" , Nucl. Instr. and Meth. B, 141, 164-168, 1998.
8. J. Davenas and X.L. Xu., "Relation between structure and electronic properties of ion irradiated polymers", Nucl. Instr. and Meth. B , 39,754,1989.
9. V.Premnath, A.Bellare, E. W. Merrill, M. Jasty and W. H. Harris, "Molecular rearrangements in ultra high molecular weight polyethylene after irradiation and long-term storage in air", Polymer, 40, 2215-2229,1999.

10. P. Bracco, del Prever E.M. Brach, M. Cannas, M.D. Luda and L. Costa, "Oxidation behaviour in prosthetic UHMWPE components sterilised with high energy radiation in a low-oxygen environment", Polymer Degradation and Stability, 91, 2030-2038, 2006.

11. E. Sokullu Urkac , A. Oztarhan, F.Tihminlioglu, N. Kaya, D. Ila, C. Muntele, S. Budak, E. Oks, A.Nikolaev, A.Ezdesir, Z. Tek, "Thermal Characterization of Ag and Ag+N ion implantated Ultra High Molecular Weight Polyethylene (UHMWPE)", Nucl. Instr. and Meth. B , 2007, In Press.

12. N. Kaya, A.Oztarhan, E. Sokullu Urkac, D. Ila, S. Budak, E. Oks, A. Nikolaev, A. Ezdesir, F.Tihminlioglu, Z. Tek, "Polymeric Thermal Analysis of C+H and C+H+Ar Ion Implanted UHMWPE Samples", Nucl. Instr. and Meth. B , 2007, In Press.

13. K. G. Kostov, M.Ueda, I. H. Tan , N.F. Leite, A.F. Beloto and G.F. Gomes, "Structural effect of nitrogen plasma-based ion implantation on ultra-high molecular weight polyethylene", Surface and Coatings Technology, 186, 287-290, 2004.

14. J. Chen, F. Zhu, H. Pan, J. Cao, D. Zhu, H. Xu, Q. Cai, J. Shen, L. Chen and Z. He, "Surface modification of ion implanted ultra high molecular weight polyethylene", Nucl. Instr. and Methods B, 169, 26-30, 2000.

15. G. Lewis, "Properties of crosslinked ultra-high-molecular-weight polyethylene", Biomaterials, Vol 22, Issue 4 , pp 371-401, 2001.

16. M. P. Gispert, A.P. Serro, R.Colaço and B. Saramago, "Friction and wear mechanisms in hip prosthesis: Comparison of joint materials behaviour in several lubricants", Wear, Vol 260, Issues 1-2 , pp. 149-158, 2006.

Mater. Res. Soc. Symp. Proc. Vol. 1020 © 2007 Materials Research Society   1020-GG07-07

Irradiation Effects of Methanol Cluster Ion Beams on Solid Surfaces

Gikan Takaoka, Masakazu Kawashita, and Takeshi Okada
Ion Beam Engineering Experimental Laboratory, Kyoto University, Nishikyo-ku, Kyoto, 615-8510, Japan

ABSTRACT

In order to investigate the interactions of methanol cluster ion beams with solid surfaces, Si substrates and SiO_2 films were irradiated at different acceleration voltages. The sputtered depth increased with increase of the acceleration voltage. When the acceleration voltage was 9 kV, the sputtered depths of Si and SiO_2 at a dose of 1×10^{16} ions/cm^2 were 1497.1 nm and 147.8 nm, respectively. The selectivity between Si and SiO_2 surfaces arose from the volatility of the reaction products. Furthermore, the sputtering yield for the Si surface was approximately seven hundreds times larger than that by Ar monomer ion beams. This suggested that chemical sputtering was predominant for the methanol cluster ion irradiation. In addition, the etching and cleaning process by the methanol cluster ion irradiation was performed on the Si surfaces contaminated with a small amount of metal particles such as Au and Al. Thus, methanol cluster ion beams have unique characteristics such as surface etching and cleaning with high sputtering yield and smooth surface.

INTRODUCTION

The demand for material processing technologies for LSI devices has recently been increasing, and the development of new surface processes has been required [1-3]. For example, surface cleaning and etching processes are of much importance in semiconductor device fabrication. It is well known that the wet process using organic liquid materials has been applied to the surface treatment for solid surfaces. However, etching by organic liquid materials, such as ethanol and methanol, is not achieved even at elevated temperatures. When advanced devices in the next generation are fabricated, conventional processes for surface cleaning and etching meet the difficulties such as removable of contamination, damage-free formation, atomically flat-surface formation.

Cluster ion beams are useful tools for the investigation of the fundamentals of solid state physics, chemistry and related materials science [4-6]. High-energy-density deposition and the collective motions of the cluster atoms during impact play important roles in the surface process kinetics such as deposition, etching, cleaning or implantation. In order to overcome the limitations associated with current wet processes for surface cleaning, we have developed a new type of liquid cluster ion source, and investigated the impact processes of the liquid cluster ions on solid surfaces [7,8]. In a previous study, using ethanol cluster ions, Si surfaces were etched with high sputtering yield, and they had a lower damage and an atomically flat surface [9,10]. These features have not been obtained by the conventional wet process using alcohol.

On the other hand, methanol has a lower boiling temperature and a higher specific heat ratio than ethanol, thereby giving the possibility of larger cluster formation. Furthermore, if methanol cluster ions with large cluster size are produced, the methanol cluster ion irradiation would exhibit high sputtering yield due to the enhancement of chemical reaction rate rather than the ethanol cluster ion irradiation. This will result in the achievement of high speed etching and cleaning, and it is expecting that a throughput for the surface treatment is improved by the methanol cluster ion irradiation. In this paper, the interaction of methanol cluster ions with Si and SiO_2 surfaces is discussed based on the results such as sputtering effects. In addition, surface cleaning by removal of contaminative metal particles using the methanol cluster ion beams is described.

EXPERIMENT

Liquid materials such as methanol were introduced into a cluster source. The methanol was heated up to 130°C by a wire heater attached around the source, and the vapors of methanol were ejected through a nozzle into a vacuum region. When the vapor pressure was larger than 4500 Torr, the methanol clusters were produced by an adiabatic expansion phenomenon. The cluster size was measured by a time-of-flight (TOF) method, and it was distributed between a few hundreds and a few tens of thousands. The peak size was approximately 3000 molecules-per-cluster, which was a few times larger than that of ethanol cluster [10]. The beam intensity of cluster ions increased with increase of the vapor pressure.

The neutral clusters of methanol entered an ionizer, and they were ionized by electron bombardment. The electron voltage for ionization (Ve) was adjusted between 0 V and 300 V, and the electron current for ionization (Ie) was adjusted between 0 mA and 250 mA. The methanol cluster ions were accelerated by applying an extraction voltage to the extraction electrode. The extraction voltage (Vext) was adjusted between 0 kV and 2 kV. The extracted cluster ions were size-separated by a retarding potential method, and the size of cluster ions used was larger than 100 molecules-per-cluster. The size-separated cluster ion beams were accelerated toward a substrate, which was set on a substrate holder. The acceleration voltage (Va) was adjusted between 0 kV and 10 kV. The substrates used were Si(100) substrates. To be compared with the cluster ion irradiation on the Si surface, SiO_2 films, which were thermally grown on the Si substrates, were also irradiated by the methanol cluster ion beams. The SiO_2 film thickness was 500 nm. The background pressure around the substrate was 1×10^{-7} Torr, which was attained using a turbo-molecular pump.

RESULTS AND DISCUSSION

The sputtering process by irradiation of methanol cluster ions on Si(100) surfaces was investigated. The sputtered depth was measured by the step profiler (Veeco Instruments: DEKTAK-3173933). Figure 1 shows the dependence of sputtered depth for Si(100) substrates and SiO_2 films on acceleration voltage. The ion dose was 1×10^{16} ions/cm². As shown in the figure, the sputtered depth increases with increase of the acceleration voltage. When the acceleration voltage is 9 kV, the sputtered depth is 1497.1 nm for Si and 147.8 nm for SiO_2, respectively. The selectivity between Si and SiO_2 surfaces arises from the volatility of the reaction products and the difference in binding energy among the materials. This suggests that

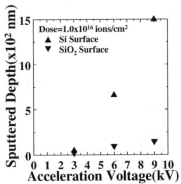

Figure 1. Dependence of sputtered depth for Si(100) substrates and SiO₂ films on acceleration voltage for methanol cluster ion beams. The electron voltage for ionization (Ve) was 200 V, and the electron current for ionization (Ie) was 200 mA. The ion dose was 1×10^{16} ions/cm².

chemical reactions between Si and methanol produce silicon hydride which is the dominant etching material for the Si surfaces. Furthermore, taking account of the sputtered depth and the ion dose, the sputtering yield was calculated by estimating the density of Si and SiO₂ such as 2.42 g/cm³ and 2.63 g/cm³, respectively. The sputtering yield of Si at an acceleration voltage of 9 kV for the methanol cluster ions was 776 atoms-per-ion, which was approximately seven hundreds times larger than that by argon (Ar) ion beam sputtering. The sputtering yield was a few times larger than that by ethanol cluster ion irradiation.

Figure 2 shows the Q-mass spectra (a) after impact of the methanol cluster ions and (b) at the atmosphere of the methanol vapors. The vacuum pressure at the atmosphere was 1×10^{-5} Torr. As shown in the figure 2(b), the fragmentation of the methanol molecule is occurred due to the ionization by an electron bombardment method in the Q-mass system. The highest peak corresponds to CH₂OH (or CH₃O) radical. To be compared with the figure 2(b), H₂ and C₂H₄ peaks for the methanol cluster irradiation become large. The dissociation of methanol molecules and the chemical reaction on the surface thereafter are described as follows;

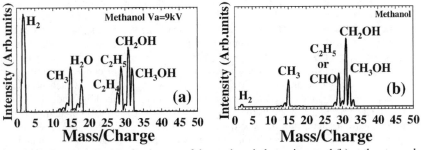

Figure 2. Q-mass spectra (a) after impact of the methanol cluster ions and (b) at the atmosphere of the methanol vapors.

$$CH_3OH \rightarrow CH_3O + H \tag{1}$$

$$H + H \rightarrow H_2\uparrow \tag{2}$$

$$Si + mH \rightarrow SiH_m\uparrow \tag{3}$$

Here, the methanol molecule after impact is dissociated to a hydrogen atom and a methoxyl radical, and they are absorbed on the Si surface. The hydrogen atom absorbed reacts with another hydrogen atom on the surface, and they form a hydrogen molecule. After staying on the surface for a while, the hydrogen molecule is desorbed from the surface. To be compared with the ethanol cluster ion irradiation, the peak of hydrogen molecule is larger, and more hydrogen atoms are produced by the methanol cluster ion irradiation.

On the other hand, the hydrogen atoms also react with Si atoms, and they can form silicon hydride, which are thereafter desorbed from the surface. These reactions are occurred immediately upon impact of methanol cluster ions on the Si surface. Furthermore, the rate of chemical reaction (ν) is described as follows;

$$\nu \propto N\frac{kT}{h} \sum_{i=1}^{n} \exp\left(-\frac{Q_i}{kT}\right) \tag{4}$$

where N is the number density of methanol molecules, h is Planck's constant, k is Boltzmann's constant, T is the temperature of methanol molecules after impact, n is the number of channels for chemical reaction, and Q_i is the activation energy for chemical reaction channel (i). According to the equation, the rate of chemical reaction increases with increasing temperature. When the methanol cluster ions impact on the Si surfaces, the cluster ions are broken up, and the multiple collisions between methanol molecules and Si atoms are occurred. As a result, many displaced Si atoms are produced, and the Si surface atoms are at more random state. This indicates that the incident energy of the cluster ions is used for heating a local region of the Si surfaces [11]. Therefore, the high temperature region formed in the cluster impact area enhances the chemical reaction between the methanol molecules and Si atoms, and the volatile materials such as silicon hydrides are formed. This result in the effective sputtering of Si surfaces by the methanol cluster ion irradiation.

The sputtered Si surfaces were measured by using an atomic force microscope (AFM). The surface roughness was 0.18 nm for the unirradiated substrate and the sputtered Si surface became rough due to the high rate sputtering. However, the surface roughness was 2.05 nm, and smooth surface with a roughness of approximately 2 nm was obtained even after sputtering. The methanol cluster ion beams have unique characteristics for surface treatment such as high sputtering yield and smooth surface, which are not achieved by the conventional wet process.

The surface cleaning of Si(100) substrates, on which metal vapors such as Au and Al were deposited previously as contaminative particles, was performed by irradiation of methanol cluster ion beams. Figure 3 shows the XPS spectra for Si(100) surfaces contaminated with (a)Au and (b)Al particles and (c)for the Si surface after irradiation of the methanol cluster ion beams, respectively. The irradiation condition was at an acceleration voltage of 9 kV and an ion dose of 1×10^{15} ions/cm^2. The amount of the contaminative metal particles was approximately 1×10^{13}

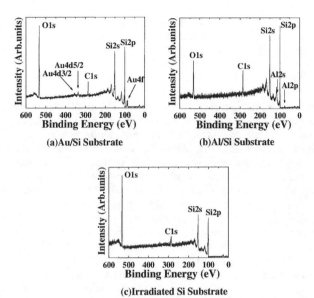

Figure 3. XPS spectra for Si(100) surfaces contaminated with (a)Au and (b)Al particles and (c)for the Si surface after irradiation of the methanol cluster ion beams. The acceleration voltage (Va) was 9 kV, and the ion dose was 1×10^{15} ions/cm^2.

atoms/cm^2. For the Si surfaces before irradiation of the cluster ion beams, XPS peaks corresponding to deposited metal particles appear together with C_{1s} and O_{1s} peaks. For all the irradiated surfaces, the metal XPS peaks disappear, and metal particles are removed by the irradiation of the methanol cluster ion beams. However, the C_{1s} and O_{1s} peaks still appear on the Si(100) surfaces, which suggest that the Si surface is covered with methanol molecules or their fragments.

The depth profile of the C_{1s} and O_{1s} peaks appearing on the Si surfaces was investigated by XPS measurement. As shown in Fig. 4, the C_{1s} and O_{1s} peaks disappear after sputtering the.

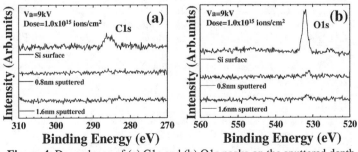

Figure 4. Dependence of (a) C1s and (b) O1s peaks on the sputtered depth.

irradiated surface by 0.8 nm. This indicates that organic materials such as methanol molecules remain on the Si surfaces after irradiation of the methanol cluster ions, but they are not implanted into the Si substrates

CONCLUSIONS

We succeeded in producing methanol clusters by an adiabatic expansion phenomenon. The intensity of the methanol clusters increased with increase of the vapor pressures, and the peak size was approximately 3000 molecules per cluster. In order to investigate the interactions of methanol cluster ion beams with solid surfaces, Si(100) substrates and SiO_2 films were irradiated at different acceleration voltages. The sputtered depth increased with increase of the acceleration voltage. When the acceleration voltage was 9 kV, the sputtered depths of Si and SiO_2 at a dose of 1×10^{16} ions/cm^2 were 1497.1 nm and 147.8 nm, respectively. The sputtering yield for the Si surface was approximately seven hundreds times larger than that by Ar monomer ion beams. This suggested that chemical reactions between Si and methanol produced silicon hydride which was the dominant etching material for the Si surfaces. In addition, the Si surfaces after sputtering had an average roughness of approximately 2 nm. Based on these results, the etching and cleaning process by the methanol cluster ion irradiation was performed on the Si surfaces contaminated with a small amount of metal particles such as Au and Al. After irradiation of the methanol cluster ions at an acceleration voltage of 9 kV and a dose larger than 1×10^{15} ions/cm^2, the contamination with the meal particles were removed from the Si surfaces. Thus, methanol cluster ion beams have unique characteristics such as surface etching and cleaning with high sputtering yield and smooth surface at an atomic level.

REFERENCES

1. C.J. Hawker and T.P. Russell, *MRS Bull.* **30** (2005) 952.
2. O. Bakajin, E. Fountain, K. Morton, S.Y. Chou, J.C. Sturm and R.H. Austin, MRS Bull. **31** (2006) 108.
3. J. W. Faul and D. Henke, *Nucl. Inst. Methods Phys. Res. B* **237**, (2005) 228.
4. I. Yamada and G. H. Takaoka, *Jpn. J. Appl. Phys.* **32** (1993) 2121.
5. K. Meinander, K. Nordlund and J. Keinonen, *Nucl. Inst. Methods Phys. Res. B* **242**, (2006) 161.
6. Y. Fujiwara, K. Kondou, Y. Teranishi, H. Nonaka, T. Fujimoto, A. Kurokawa, S. Ichimura and M. Tomita, *Surf. Interface Anal.* **38** (2006) 1539.
7. G. H. Takaoka, H. Noguchi, T. Yamamoto and T. Seki, *Jpn. J. Appl. Phys.* **42**, (2003) L1032.
8. G. H. Takaoka, H. Noguchi, K. Nakayama, Y. Hironaka and M. Kawashita, *Nucl. Inst. Methods Phys. Res. B* **237** (2005) 402.
9. G. H. Takaoka, H. Noguchi and M. Kawashita, *Nucl. Inst. Methods Phys. Res. B* **242** (2006) 417.
10. G. H. Takaoka, K. Nakayama, T. Okada and M. Kawashita, *Proc. 16th Int. Conf. on Ion Implant. Technol.* (CP866, AIP, 2006) p.321.
11. Z. Insepov and I. Yamada, *Surf. Rev. Lett.* **3**, (1996) 1023.

Mater. Res. Soc. Symp. Proc. Vol. 1020 © 2007 Materials Research Society 1020-GG07-08

Nano- and Micro-Structural Evolution of UHMWPE by Ion Beam

F. Calzzani[1], B. Chhay[1], R. Zimmerman[1], A. Oztarhan[2], and D. ILA[1]

[1]Center for Irradiation of Materials, Alabama A&M University, 3900 Meridian Street, Normal, AL, 35762
[2]Bioengineering, Ege University, Izmir Bornova, 35100, Turkey

ABSTRACT

It is important to produce uniform nano-patterns with no possibility of surface exfoliation on polyethylene devices used in medical and in aerospace industry. We studied the change in the surface morphology of polyethylene at nanoscale using MeV ion beam. We have investigated the change in the surface morphology before and after ion bombardment. We have made an attempt to change the morphology to produce a uniform surface with reduced cracks and reduced granularity. For this process we have chosen ultra-high-molecular-weight polyethylene (UHMWPE). Coupons of these materials were exposed to various fluences of MeV Ag^+ ions. The surface morphology and the change in the chemical structure were studied using scanning micro Raman, FTIR and AFM.

1. INTRODUCTION

Joint replacements are a very common surgical procedure used to restore function to cases that joints have been afflicted by major trauma or degenerating disease. Usually, those surgical approaches for joint replacement use a metal surface articulating against a component of UHMWPE. The use of UHMWPE became popular in total joint replacements due to its properties including biocompatibility, wear, friction, ductility and impact load resistance. However, wear particle-induced aseptic loosening of joint replacement prostheses remains a major cause of revision surgeries for the commonly used metal/ultra-high molecular weight polyethylene (UHMWPE) [1].

In addition to surface wear problems and despite the efforts of countless researchers to combat it, device-associated infection remains a major problem in medical care. Infection at medical replacements and indwelling catheters, for example, can result from contaminated disinfectants, from the hands of medical personnel, or as a result of self infection from a patient's own microflora. Such infections are not easily treated, since proliferating bacteria on the surface of the medical material can secrete a polysaccharide biofilm or "slime" difficult for systemic antibiotics to penetrate. Irradiation with Gamma rays [2] have been used extensively as the standard method of disinfection, however it's well known that this promotes undesirable oxidation under the prosthesis surface raising the wear production and causing a artificial aging.

One alternative way of addressing device- related infection is to incorporate antimicrobial agents directly onto the surface of the device. Silver compounds (silver chloride or silver oxide) are a popular choice for infection-resistant coatings, but many commercially available silver-coated materials are of marginal effectiveness because the hydrophobic polymer matrix limits the silver ion concentration near the device surface.

In order to induce surface modifications in wear properties as well as in antimicrobial properties we use modification by silver ion bombardment [3]. Materials modification by ion-beam process has recently become an interesting topic in the field of surface engineering. We have used such methods to change surface properties creating layers with more suitable resistance to friction and in addition we have used silver ion bombardment to create a thin layer resistant to infection.

2. EXPERIMENTAL

We used a Pelletron accelerator located in the Center for Irradiation of Materials (CIM) at Alabama A&M University (AAMU) to do 1 MeV Ag+ ions bombardment in six samples of UHMWPE at fluences from 10^{13} to 10^{15} ions/cm^2.

Before the bombardment, 500 μm thick UHMWPE sheets had their surfaces smoothed by pressing them between two glass surfaces under 70 °C heat treatment. This process "printed" the smoothed surface of glass over the UHMWPE surface. In order to evaluate the damage in specific positions of the samples after ion bombardment, a copper grating with 500 square/mm was pressed in the film at room temperature through a glass plate, forming 20 μm×20 μm well defined square patterns as shown in Atomic Force Microscopy image (Fig 1).

Before and after ion bombardment, the near-surface of the UHMWPE was investigated by Atomic Force Microscopy performed in a SOLVER P47H Scanning Probe Microscope (NT-MDT, NTI Instruments) in air environment, at room temperature. Measurements were made in the non-contact/tapping mode using standard silicon cantilevers (NSG11) with nominal spring constant kc=2.0N/m, cone angle 22°, typical curvature radius 10nm and typical resonance 150kHz. During the analysis, the tips were positioned inside the limits defined by the printed squares in the polymer surface to allow a comparison of the imaged structures before and after bombardment.

Additional chemical structure analysis was performed by optical spectroscopy methods. Optical absorption spectrometry measurements were made using a Cary 500 spectrophotometer. ATR-FTIR measurements were made using a Tensor 27 spectrophotometer with a Pike-ATR accessory, equipped with a ZnSe crystal. This technique is surface sensitive and suitable for ion implanted analysis in films where the damage is located at few microns in depth in the material surface. Micro-Raman analyses were made in a Lab-Ram equipped with a He-Ne laser (637.18 nm).

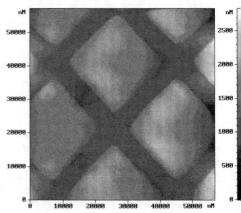

Fig 1. AFM picture of gratings printed in UHMWPE surface after surface smoothing process using glass

3. RESULTS AND DISCUSSION

After ion bombardment we notice a gradual increase of the samples opacity for accumulated fluences, changing from white semi–transparent in virgin non-irradiated samples till dark brown opaque obtained in samples irradiated at 10^{15} ions/cm^2. Optical measurements of the bombarded UHMWPE samples show a shift from 250 nm to 500 nm (Fig. 2). The optical absorption spectra show no characteristic absorption peak of silver and indicate no formation of silver clusters. The color change of UHMWPE samples from white to brown indicates the formation of carbon double bounds and conjugated carbon bonds that absorb the blue light. The apparent absorption is partly a result of light scattered by surface roughness induced by the bombardment.

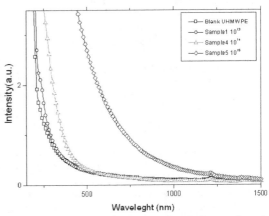

Fig 2. Optical absorption spectra for virgin sample and three different fluences of silver ion bombardment for 10^{13} ions/cm^2, 10^{15} ions/cm^2 and 10^{14} ions/cm^2.

ATR-FTIR spectroscopy was performed at the surface UHMWPE and it shows the typical spectral results of polyethylene. Peaks at 2916 cm^{-1} to 2855 cm^{-1}, 1492 cm^{-1} to 1466 cm^{-1} and 749 cm^{-1} to 723 cm^{-1} are attributed to C-H and C-C absorption in IR Region, respectively.

After bombardment, the spectra shown in Fig. 3 display additional absorption broad bands at 1740 cm^{-1} and 1634 cm^{-1} showing the evolution of C=O and C=C bonds. The broad band around 1600 cm^{-1} also suggests the presence of the aromatic and/or olefinic structures in bombarded samples. Another broad band around 1100 cm^{-1} arises, which indicates the formation of carbonyls and byproducts associated with oxidation [4, 5]. This is attributed to C–O stretching vibration resulting from the formation of hydroperoxides following the exposure to residual air in the implantation chamber. This bond is a consequence of hydrogen leaving the polymer and allowing oxygen to attach to the dangling bonds. Those bands increase gradually till fluences of 10^{14}ions/cm^2 and decrease to samples bombarded at 10^{15}ions/cm^2.

Simulations performed using SRIM 2006 show for 1MeV silver (Ag$^+$) in UHMWPE a longitudinal range of 0.71 ± 0.14 µm and approximately 6300 vacancies produced per incident ion. When 1MeV silver ions enter the UHMWPE, they loose energy in a sequence of elastic and inelastic collisions resulting in excitations of bound electrons of the medium, excitations of the electron cloud of the ion, vacancies and positions changes by directly atom/atom interaction. Such reactions break long chains of polyethylene on the surface and also create cross-links between consecutive planes [6, 7]. Atomic Force Microscopy (AFM) was accomplished in order to see any surface modifications due to ion bombardment.

AFM images obtained previously were compared to images of samples after silver bombardment. One can notice that the printed grating retains its shapes even at the 10^{15} ions/cm^2 highest fluence, making possible a comparison with the same area of the samples before and after bombardment.

The sequence of AFM images from Fig. 3 show the evolution of surface modifications induced by different silver fluences comparing with virgin sample. There is a slightly change from the virgin surface to the surface treated with 10^{14} ions/cm^2. Micro modifications become clear in AFM images for samples treated with fluence of 10^{15} ions/cm^2. Such modifications are revealed in nano-scale images that show a progressive modification in the polyethylene chains due to formation of amorphous carbon surface layer that produces an enhancement of micro-hardness [10].

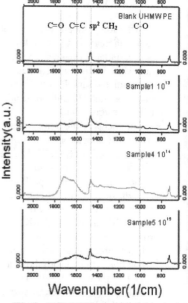

Fig 3. ATR-FTR spectra of (a) virgin sample and three fluences of silver ion bombardment (b) 10^{13} ions/cm^2, (c) 10^{15} ions/cm^2 and (d) 10^{14} ions/cm^2.

The Raman spectra of virgin and bombarded samples are shown in Fig. 5. In contrast with FTIR the RAMAN spectra show, with the exception of photoluminescence, no peak differences in samples bombarded at 10^{12} ions/cm^2 to 10^{14} ions/cm^2 fluences and the virgin. Clearly spectral changes occur for 10^{15} ions/cm^2 fluence. From Fig. 5 (a) one can notice that the UHMWPE spectral intensity increase significantly for samples treated with 10^{14} ions/cm^2 when compared with samples bombarded at lower fluences. This effect can be attributed to two phenomena that will be further investigated in later analysis. Even though the optical absorption shows no evidence of silver clustering, the first suggested effect is the Surface Enhancement Raman Scattering (SERS) that is observed for molecules found close to silver nanoparticles because of surface plasmon resonance.

The second possible effect suggested is the increase of fluorescence caused by implantation, which can be ascribed to an increase of defects in the polymer crystal caused by the implantation of the energetic ions. At 10^{15} ions/cm^2 the bombarded samples show significant structural changes. Previous well-defined peaks in 1070 cm^{-1}, 1120 cm^{-1}, 1300 cm^{-1} and 1400-1500 cm^{-1} vanish, raising broad peaks.

The main features of bombarded sample spectra by fluence 10^{15} ions/cm^2 shown in Fig. 5 (b) are some partially overlapping broad bands. The peaks before bombardment become spread in the bombarded sample and still have spectral contribution in non-defined bands centered at 1116 cm^{-1}, 1220 cm^{-1} and 1300 cm^{-1}. The highest band is centered at approximately 1542 cm^{-1} and it's related to G-Band associated with sp2 graphite carbon.

It is well known that there is only a sharp G band in the Raman spectrum of single crystal graphite, which originates from phonons at the Brillouin zone center that satisfy the momentum selection rules at k=0. Thus the appearance of any other bands in the Raman spectra of carbon materials is caused by the breakdown of the selection rules due to the disorder in the structure of samples. The position, intensity, and width of these bands depend on the degree of disorder in the samples. The physical properties of amorphous carbon have been the subject of intense experimental and theoretical work [8]. Amorphous carbon is a disordered phase of carbon without long-range order containing carbon atoms, mostly in graphite-like sp2 and diamond-like sp3 hybridization states, and its physical properties depend strongly on the sp2/sp3 ratio.

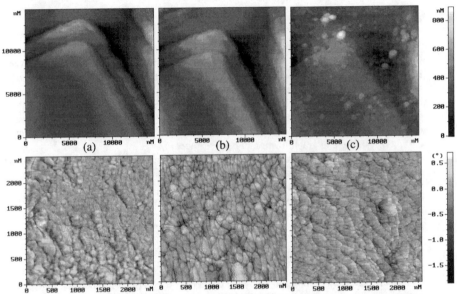

Fig 4. Evolution of UHMWPE surface by silver ion bombardment in two different scales showing (top) micro- and (bottom) nano- structures for (a) virgin sample, (b) fluence 10e14 ions/cm2 and (c) fluence 10e15 ions/cm2.

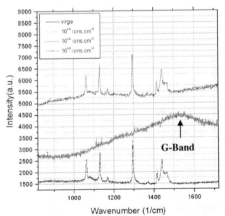

Fig 5. Micro RAMAN spectra of virgin sample and three Silver ion bombarded samples by fluences 10^{13} ions/cm^2, 10^{14} ions/cm^2 and 10^{15} ions/cm^2.

There are many forms of sp2-bonded carbons with various degrees of graphitic ordering, ranging from microcrystalline graphite to glassy carbon.

Both methods, optical absorption and FTIR, suggest evidence of the presence of C=C and C=O formation due to dehydrogenation in all bombarded samples. For 10^{15} ions/cm^2 samples, the RAMAN results show clearly the increasing of G-Band peak and the spreading and vanishing of the original peaks. All optical methods show evidence of amorphous carbon and graphitization present near the surface. In addition, the AFM images show the surface evolution of UHMWPE polymer chains due to Ag$^+$ ion bombardment for different fluences.

Generally, for a given ion species, the hardening increases with the fluence until saturation, which is a function of the ion energy [9] and of the complexity of polymeric structure. It has been shown that the sputtered amorphous carbon films with sp2-bonding dominated structure (graphite-like) also exhibited super hardness, excellent tribological properties, and high load-bearing capacity [10].

4. CONCLUSIONS

The optical measurements performed on the bombarded UHMWPE reveal low damage levels in fluences below 10^{14} ions/cm^2 and little or no evidence of any chemical reaction between the implanted species and the host material. There is a certain amount of carbonization and graphitization following higher dose implantation (10^{15} ions/cm^2) as well as formation of hydroperoxides following dehydrogenation of the polymer.

We studied the surface morphology evolution using AFM that showed us a progressive rupture of polymeric chains in function of applied fluences due to dehydrogenation, amorphous carbon increasing and graphitization in near-surface.

All optical photospectrometric methods show evidences of an increase of amorphous carbon and graphitization near the surface of samples, a desirable characteristic to improve mechanical and tribological properties of UHMWPE samples.

Since the controlled increase of micro-hardness is a suitable characteristic to improve the surface mechanical properties in our samples, further investigations will be undertaken in order to optimize and quantify the surface evolution in UHMWPE samples.

5. ACKNOWLEDGEMENTS

This research was sponsored by by the Center for Irradiation of Materials, Alabama A&M University and supported by NSF-EPSCoR grant No. EPS-0447675.

6. REFERENCES

[1] Kurtz SM, Muratoglu OK, Evans M, Edidin AA. Advances in the processing, sterilization, and crosslinking of ultra-high molecular weight polyethylene for total joint arthroplasty. Biomaterials 1999; 20:1659–88.
[2]. Sutula L, Collier J, et al. The Otto Aufranc Award. Impact of gamma sterilization on clinical performance of polyethylene in the hip. Clin Orthop 1995:28.
[3]. Onate JI, Alonso F, Garcia A: Improvement of tribological properties by ion implantation. Thin Solid Films 317:471, 1998
[4] R.A. Minamisawa, et al., Effects of low and high energy ion bombardment on ETFE polymer, Nucl. Instr. And Meth. B (2007), doi:10.1016/j.nimb.2007.01.226
[5] M.A. Parada and A. De Almeida, Nucl. Instr. and Meth. B 241 (2005) , p. 521-525.
[6]. Muratoglu OK, Bragdon CR, et al. Unified wear model for highly crosslinked ultra-high molecular weight polyethylenes (UHMWPE). Biomaterials 1999;20:1463.
[7]. Muratoglu OK, Merrill EW, et al. Effect of radiation, heat, and aging on in vitro wear resistance of polyethylene. Clin Orthop Relat Res 2003;417:253.
[8]. A. C. Ferrari and J. Robertson, "Interpretation of Raman spectra of disordered and amorphous carbon", Phys. Rev., B 61, (2000) 14095.
[9]. E.H. Lee, G.R. Rao, M.B. Lewis and L.K. Mansur. J. Mater. Res. 9 (1994), p. 1043
[10]. V. Kulikovsky, K. Metlov, A. Kurdyumov, P. Bohac and L. Jastrabik. Diamond Relat. Mater. 11 (2002), p. 1467

Mater. Res. Soc. Symp. Proc. Vol. 1020 © 2007 Materials Research Society   1020-GG07-09

MeV Ion-Beam Bombardment Effects on the Thermoelectric Figures of Merit of Zn4Sb3 and ZrNiSn-Based Half-Heusler Compounds

S. Budak[1], S. Guner[1], C. Muntele[1], C. C. Smith[2], B. Zheng[1], R. L. Zimmerman[1], and D. ILA[1]

[1]Center for Irradiation of Materials, Alabama A&M University, 4900 Meridian Street, PO Box 1447, Normal, AL, 35762
[2]MSFC, NASA, MSFC, Huntsville, AL, 35812

Abstract

Semiconducting β-Zn4Sb3 and ZrNiSn-based half-heusler compound thin films were prepared by co-evaporation for the application of thermoelectric (TE) materials. High-purity solid zinc and antimony were evaporated by electron beam to grow the β-Zn4Sb3 thin film while high-purity zirconium powder and nickel tin powders were evaporated by electron beam to grow the ZrNiSn-based half-heusler compound thin film. Rutherford backscattering spectrometry (RBS) was used to analyze the composition and thickness of the thin films. The grown thin films were subjected to 5 MeV Si ions bombardment for generation of nanostructures in the films. We measured the thermal conductivity, Seebeck coefficient, and electrical conductivity of these two systems before and after 5 MeV Si ions beam bombardment. The two material systems have been identified as promising TE materials for the application of thermal-to-electrical energy conversion, but the efficiency still limits their applications. The electronic energy deposited due to ionization in the track of MeV ion beam can cause localized crystallization. The nanostructures produced by MeV ion beam can cause significant change in both the electrical and the thermal conductivity of thin films, thereby improving the efficiency. We used the 3ω-method measurement system to measure the cross-plane thermal conductivity ,the Van der Pauw measurement system to measure the electrical conductivity, and the Seebeck-coefficient measurement system to measure the cross-plane Seebeck coefficient. The thermoelectric figures of merit of the two material systems were then derived by calculations using the measurement results. The MeV ion-beam bombardment was found to decrease the thermal conductivity of thin films and increase the efficiency of thermal-to-electrical energy conversion.

Keywords: Ion bombardment, Thermoelectric properties, Rutherford backscattering spectrometry, Van der Pauw method, 3w method, Seebeck coefficient, Figure of merit

*Corresponding author: D. ILA; Tel.: 256-372-5866; Fax: 256-372-5868; Email: ila@cim.aamu.edu

1. INTRODUCTION

Thermoelectric materials are very important due to the interest in their applications in thermoelectric power generation and microelectronic cooling [1]. Thermoelectric power generation could convert heat to electricity directly. The ZrNiSn half-Heusler alloy is one of the potential candidates for the thermoelectric materials and has recently received great interest [2]. β-Zn$_4$Sb$_3$ with a complex hexagonal crystal structure has also been discovered to be one of the promising candidates for thermoelectric power generation applications [3]. Effectiveness of the thermoelectric materials depend on a low thermal conductivity and a high electrical conductivity [4]. The performance of the thermoelectric materials and devices is described by a dimensionless Figure of Merit, $ZT = S^2 \sigma T / \kappa$, where S is the Seebeck coefficient, σ is the electrical conductivity, T is the absolute temperature, and κ is the thermal conductivity [5]. Higher ZT can be reached by increasing S, increasing σ, or decreasing κ.

Since the bulk form of the half-heusler β-Zn$_4$Sb$_3$ has a higher figure of merits at higher temperatures [6] and the half-heusler ZrNiSn has been studied by many researchers for the candidate as the practical thermoelectric materials because of their good thermoelectric performance and low toxicity of the constituent elements [7], we worked on the thin films of these materials. In this study, we report on the growth of half-Heusler alloys of Zn$_4$Sb$_3$ and ZrNiSn thin films on the silica substrates using an ion-beam assisted deposition (IBAD) system, and high energy Si ions bombardments of the thin films for reducing thermal conductivity and increasing electrical conductivity.

2. EXPERIMENTAL

Semiconducting β-Zn$_4$Sb$_3$ and ZrNiSn-based half-heusler compound thin films on the silica substrates were grown with the Ion Beam Assisted Deposition (IBAD) system. High-purity solid zinc and antimony were evaporated by electron beam to grow the β-Zn$_4$Sb$_3$ thin film while high-purity zirconium powder and nickel tin powders were evaporated by electron beam to grow the ZrNiSn-based half-heusler compound thin film. The thickness of the films were controlled by an INFICON deposition monitor. The film geometries used in this study are shown in Fig.1. Fig.1a shows the geometry of Zn$_4$Sb$_3$ thin film from the cross-section while Fig.1b shows the geometry of ZrNiSn thin film. The geometries in Fig.1 show two Au contacts on the top and the bottom of the films. These two Au contacts were used in the Seebeck coefficient measurement.

The electrical conductivity was measured by the Van der Pauw system and the thermal conductivity was measured by the 3ω technique. The electrical conductivity, the thermal conductivity and the Seebeck measurements were performed at a room temperature of $22\,^{\circ}C$. One could find detailed information about 3ω technique in Refs. [8-10]. The 5 MeV Si ions bombardments were used by the Pelletron ion beam accelerator at the Alabama A&M University's Center for Irradiation of Materials (AAMU-CIM).

The fluences used for the bombardments were $1x10^{14}\,ions/cm^2$, $5x10^{14}\,ions/cm^2$ and $1x10^{15}\,ions/cm^2$. Rutherford backscattering spectrometry (RBS) measurement was performed using 2.1 MeV He$^+$ ions in an IBM scattering geometry with the particle detector placed at 170 deg from the incident beam to monitor the film thickness and stoichiometry before and after 5 MeV Si ions bombardments [11, 12].

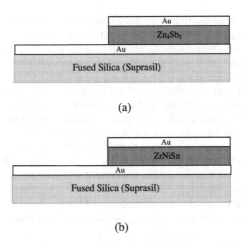

(a)

(b)

Fig. 1. Geometry of sample from the cross-section.

3. RESULTS AND DISCUSSION

Fig. 2a. shows RBS spectrum of Zn_4Sb_3 thin film on Glassy Polymeric Carbon (GPC) substrate when the sample is at the normal angle. RUMP simulation [13] gives information about the amount of the elements in the deposited samples and the thickness of the deposited films. According to the RUMP simulation for Zn_4Sb_3 thin film, the ratio between Zn and Sb was found as 4:3, and the thickness of the Zn_4Sb_3 was found as 680 nm. Fig. 2b. shows RBS spectrum of ZrNiSn thin film on Glassy Polymeric Carbon (GPC) substrate when the sample is at the normal angle. The RUMP simulation for ZrNiSn gave the thickness of 110 nm and the ratio among the Zr, Ni, and Sn as 1:1:1, respectively.

Fig. 3 shows the thermoelectric properties of Zn_4Sb_3 and ZrNiSn thin films. Fig. 3a shows the square of the Seebeck coefficient change for Zn_4Sb_3 thin film depending on the fluences of the bombardments. As seen from Fig.3a, the square of the Seebeck coefficient for Zn_4Sb_3 thin film tends to decrease starting from the virgin case of the Zn_4Sb_3 thin film to the fluence of $5x10^{14} ions / cm^2$. After the fluence of $5x10^{14} ions / cm^2$, the square of the Seebeck coefficients started to increase. The requirement of a high Seebeck coefficient is natural since S is a measure of the average thermal energy which is carried per charge (electron or hole) [14]. Fig. 3b shows the electrical conductivity change for Zn_4Sb_3 thin film depending on the fluences. As seen from Fig. 3b, the electrical conductivity for Zn_4Sb_3 thin film started to increase when the first bombardment of $1x10^{14} ions / cm^2$ was applied. After the fluence of $1x10^{14} ions / cm^2$, the electrical conductivity for Zn_4Sb_3 thin film started to decrease until the fluence of

$5x10^{14} ions/cm^2$. After the fluence of $5x10^{14} ions/cm^2$, the electrical conductivity for Zn$_4$Sb$_3$ thin film started to increase. The fluences of $1x10^{14} ions/cm^2$ and $5x10^{14} ions/cm^2$ behaved as turning points for the electrical conductivity for Zn$_4$Sb$_3$ thin film. This shows that ion bombardment caused an increase in the electrical conductivity until one certain fluence was reached. While the virgin sample is being bombarded with the 5 MeV Si ions, the numbers of the charge carriers in both the conduction and valence bands increase. This increase causes shorter energy gap between the conduction and valence bands. The shorter energy gap causes increase in the electrical conductivity. The decrease in the electrical conductivity might be due to the degenarations in both the conduction and the valence bands. The increase in the electrical conductivity is one of the desired conditions for both the thermoelectric materials and the devices. Fig. 3c shows the thermal conductivity change for Zn$_4$Sb$_3$ thin film depending on the fluences. As seen from Fig. 3c, the thermal conductivity for Zn$_4$Sb$_3$ thin film decreases starting from the virgin case to the fluence of $5x10^{14} ions/cm^2$. The decrease in the thermal conductivity is the another desired property in both the thermoelectric materials and devices. After the fluence of $5x10^{14} ions/cm^2$, the thermal conductivity for Zn$_4$Sb$_3$ thin film increases. The high energy ion bombardment can produce nanostructures and modify the property of thin films [15], resulting in lower thermal conductivity and higher electrical conductivity. Fig. 3d shows the fluence dependence of figure of merit of Zn$_4$Sb$_3$ thin film. As seen from Fig. 3d, the figure of merit value started to increase when the 5 MeV Si ions bombardments were introduced until the fluence of $1x10^{14} ions/cm^2$. After the fluence of $1x10^{14} ions/cm^2$, the figure of merit started to decrease. Both the good thermoelectric materials and devices should have higher figure of merit. The Figure of Merit for Zn$_4$Sb$_3$ thin film increases from 0.176 at zero fluence to 0.53 at $1x10^{14} ions/cm^2$ fluence. After the fluence of $1x10^{14} ions/cm^2$, the figure of merit decreases to 0.032 at $1x10^{15} ions/cm^2$ fluence.

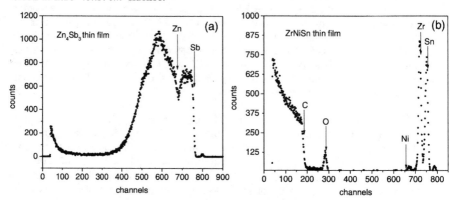

Fig. 2. He RBS spectra of Zn$_4$Sb$_3$ and ZrNiSn films on GPC substrate.

Fig. 3e shows the square of the Seebeck coefficient change for ZrNiSn thin film depending on the fluences of the bombardments. As seen from Fig.3e, the square of the Seebeck coefficient for ZrNiSn thin film tends to increase starting from the virgin case of the ZrNiSn thin film to the fluence of $1x10^{14} ions/cm^2$. After the fluence of $1x10^{14} ions/cm^2$, the square of the Seebeck coefficient started to decrease until the fluence of $5x10^{14} ions/cm^2$. After the fluence of $5x10^{14} ions/cm^2$, the square of the Seebeck coefficient for ZrNiSn thin film increases. Fig. 3f shows the electrical conductivity change for ZrNiSn thin film depending on the fluences. As seen from Fig. 3f, the electrical conductivity for ZrNiSn thin film decreased when the first bombardment of $1x10^{14} ions/cm^2$ was applied. After the fluence of $1x10^{14} ions/cm^2$, the electrical conductivity for ZrNiSn thin film kept constant at around zero. The decrease in the electrical conductivity might be due to the degenerations in both the conduction and the valence bands. Fig. 3g shows the thermal conductivity change for ZrNiSn thin film depending on the fluences. As seen from Fig. 3g, the thermal conductivity for ZrNiSn thin film increases starting from the virgin case to the fluence of $5x10^{14} ions/cm^2$. After the fluence of $5x10^{14} ions/cm^2$, the thermal conductivity for ZrNiSn thin film decreases. Fig. 3h shows the fluence dependence of figure of merit of ZrNiSn thin film. As seen from Fig. 3h, the figure of merit value started to decrease when the 5 MeV Si ions bombardments were introduced until the fluence of $5x10^{14} ions/cm^2$. After the fluence of $5x10^{14} ions/cm^2$, the figure of merit kept constant at around $5.93x10^{-7}$. The Figure of Merit for ZrNiSn thin film decreases from $9.11x10^{-4}$ at zero fluence to $5.93x10^{-7}$ at $1x10^{15} ions/cm^2$ fluence.

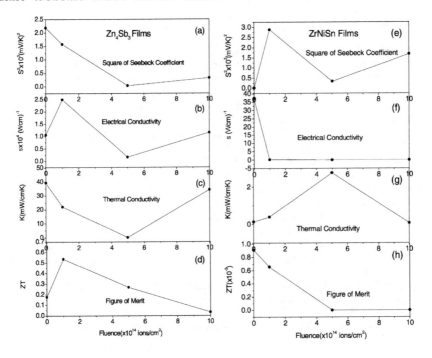

Fig.3. Thermoelectric properties of Zn4Sb3 and ZrNiSn thin films.

4. CONCLUSION

We have grown semiconducting β-Zn_4Sb_3 and ZrNiSn-based half-heusler compound thin films on the silica substrates using IBAD system. Rutherford backscattering spectrometry (RBS) was used to analyze the composition of the thin films. The thin films were then bombarded by 5 MeV Si ions for generation of nanostructures in the films. As we explained in the text, the 5 MeV Si ion bombardments caused increment in the electrical conductivity, decrement in the thermal conductivity, and thus increment in the figure of merit when the suitable fluences of bombardment were chosen. We got a figure of merit of 0.53 for Zn_4Sb_3 thin film at the fluence of $1x10^{14} ions/cm^2$. We will continue to get a higher figure of merit for these samples.

Acknowledgement

Research sponsored by the Center for Irradiation of Materials, Alabama A&M University and by the AAMURI Center for Advanced Propulsion Materials under the contract number NNM06AA12A from NASA, and by National Science Foundation under Grant No. EPS-0447675.

References

1. S. Budak, C. C. Smith, B. Zheng, C. I. Muntele, R. L. Zimmerman and D. Ila, Mater. Res. Soc. Symp. Proc. Vol.974 © 2007 Materials Research Society 0974-CC10-16.
2. X. Y. Huang, Z. Xu, L. D. Chen, Solid State Communications 130,181 (2004).
3. Kazuhiro Ito, Lanting Zhang, Katsuyuki Adachi and Masaharu Yamaguchi, Mater. Res. Soc. Symp. Proc. Vol.793 © 2004 Materials Research Society S5.1.
4. Brian C. Scales, Science, 295,1248 (2002).
5. B. Y. Yoo, C. -K. Huang, J. R. Lim, J. Herman, M. A. Ryan, J. -P. Fleural, N. V. Myung, Electrochemical Acta 50, 4371 (2005).
6. Terry M. Tritt and M.A. Subramanian, MRS Bulletin, 31,190 (2006).
7. Shigeru Katsuyama, Ryosuke Matsuo, Mikio Ito, J. Alloys Compd. 428 ,262 (2007).
8. L. R. Holland, R. C. Simith, J. Apl. Phys. 37,4528 (1966).
9. D. G. Cahill, M. Katiyar, J. R. Abelson, Phys. Rev.B 50,6077 (1994).
10. T. B. Tasciuc, A.R. Kumar, G. Chen, Rev. Sci. Instrum. 72 ,2139 (2001).
11. J. F. Ziegler, J. P. Biersack, U. Littmark, The Stopping Range of Ions in solids, Pergamon Press, Newyork, 1985.
12. W. K. Chu, J. W. Mayer, M. -A. Nicolet, Backscattering Spectrometry, Academic Press, New York, 1978.
13. L. R. Doolittle, M. O. Thompson, RUMP, Computer Graphics Service, 2002.
14. G. Chen, A. Narayanaswamy, C.Dames, Superlattices and Microstructures 35 (2004) 161.
15. S. Budak, B. Zheng, C. Muntele, Z. Xiao, I. Muntele, B. Chhay, R. L. Zimmerman,. L. R Holland, and D. Ila, Mater. Res. Soc. Symp. Proc. Vol. 929 © 2006 Materials Research Society 0929-II04-10.

Mater. Res. Soc. Symp. Proc. Vol. 1020 © 2007 Materials Research Society   1020-GG07-10

In Vitro Apatite Formation on Polymer Substrates Irradiated by the Simultaneous Use of Oxygen Cluster and Monomer Ion Beams

Masakazu Kawashita, Rei Araki, and Gikan H Takaoka
Ion Beam Engineering Experimental Laboratory, Kyoto University, Nishikyo-ku, Kyoto, 615-8510, Japan

ABSTRACT

Polyethylene (PE) and silicone rubber substrates were irradiated at an acceleration voltage of 7 kV and a dose of 1×10^{15} ions/cm^2 by the simultaneous use of oxygen cluster and monomer ion beams, and then soaked in CaCl$_2$ solution. Apatite-forming ability of the substrates was examined using a metastable calcium phosphate solution that had 1.5 times the ion concentrations of a normal simulated body fluid (1.5SBF). After the irradiation, the hydrophilic functional groups such as COOH and silicon oxide clusters (SiO$_x$) were formed at the PE and silicone rubber surfaces, respectively. The hydrophilicity of the substrates was remarkably improved by the irradiation. The irradiated PE and silicone rubber substrates formed apatite in 1.5SBF, whereas unirradiated ones did not form it. These results suggest that the functional groups such as COOH groups and Si–OH groups induced apatite nucleation in 1.5SBF.

INTRODUCTION

Artificial ligaments which are two-dimensional polyethylene terephthalate (PET) fabrics have been used for reconstruction of damaged ligaments. In the reconstruction, the artificial ligaments are fixed to bone by using bone plug and staples. However, a long-term is often required for complete cure, because PET does not bond to living bone directly but bond to the bone through fibrous tissues and hence the fixation of the artificial ligaments becomes sometimes unstable. If the surface of PET is modified to show bone-bonding ability (i.e. bioactivity), it is believed that the term for complete cure can be shortened remarkably. In the case of indwelling catheters of silicone rubber, serious infection sometimes occurred owing to its poor biocompatibility. It is believed that such infection can be suppressed by apatite coating on the silicone rubber surface, since it is well known that apatite shows an excellent biocompatibility. If the surface of silicone rubber is modified to show bioactivity, apatite can be coated on the silicone rubber surface by soaking in aqueous solution supersatuared with respect to apatite such as simulated body fluid (SBF).

In order to obtain bioactive polymers, we should form functional groups such as Si–OH and COOH groups effective for apatite nucleation, on the surface of polymers [1]. Various attempts to modify the polymer surface with the bioactive phase have been made [2,3], but it is difficult to form the bioactive phase stably on the surfaces of hydrophobic polymers such as PE and silicone rubber.

The ion beam process has a high potential for surface modification of polymers. The predominant properties of the ion beam process are based on the ability to control the kinetic energy precisely by adjusting the acceleration energy. In addition, atomic, molecular and cluster ions are available, and the interaction of these ions with the substrate surface can be greatly varied depending on the ion species used. The cluster is an aggregate of a few tens to several thousands of atoms or molecules, and constitutes a new phase of matter, with significantly different properties than solids, liquids and gases [4]. The effect of cluster ion impact on a solid surface is much different from the single ion effect, and shows the high-density irradiation and the multiple collision. Another feature of the cluster ion impact is the low-energy irradiation, and the damage-free irradiation can be performed by adjusting the incident energy of the cluster ion beam [5]. On the other hand, the monomer ion beams with high incident energy produce free radicals and bond scission on the irradiated polymer surfaces that are effective for cross-linking and/or carbonization [6]. In this study, the PE and silicone rubber substrates were irradiated with O_2 cluster and monomer ion beams, and in vitro the apatite-forming ability of the irradiated PE and silicone rubber substrates was investigated in a metastable calcium phosphate solution (1.5SBF) that had 1.5 times the ion concentrations of a normal simulated body fluid.

EXPERIMENT

Simultaneous irradiation of oxygen cluster and monomer (O_2 CM) ion beams

Details of the experimental apparatus for cluster ion beams have been given elsewhere [6]. Briefly, neutral beams of O_2 clusters were produced by adiabatic expansion of a high pressure gas through a glass nozzle into a high vacuum chamber. In this study, we used O_2 cluster and monomer (O_2 CM) beams, since our previous study [7] suggests O_2 CM ion beams are more effective for surface modification of polymers than O_2 cluster ion beams or O_2 monomer ion beams. The O_2 CM ion beams were introduced into the ionization and irradiation chamber under the inlet gas pressure of 4.0×10^4 Pa and ionized by electron bombardment with an ionization voltage of 300 V and an electron current of 300 mA. PE and silicone rubber (polydimethylsiloxane; PDMS) substrates $20 \times 20 \times 1$ mm^3 in size were irradiated at an ion dose of 1×10^{15} ions/cm^2 by the simultaneous use of O_2 CM ion beams. The acceleration voltage (V_a) for the ion beams was set at 7 kV.

Treatment in CaCl$_2$ aqueous solution and soaking in 1.5SBF

The substrates were soaked for 24 h at 36.5 or 42 °C in 30 mL of 1 mol/dm^3 CaCl$_2$ aqueous solution in order to enhance apatite nucleation, and then immersed for 7 days at 36.5 or 42 °C in 30 mL of a metastable calcium phosphate solution (1.5SBF) that had 1.5 times the ion concentrations of a normal simulated body fluid. The 1.5SBF was prepared by dissolving NaCl, NaHCO$_3$, KCl, K$_2$HPO$_4 \cdot$3H$_2$O, MgCl$_2 \cdot$6H$_2$O, CaCl$_2$ and Na$_2$SO$_4$ in ultrapure water, and buffered at pH 7.40 at 36.5 · C with tris(hydroxymethyl)aminomethane (($CH_2OH)_3CNH_2$) and 1 mol/dm^3 HCl aqueous solution. The ion concentrations and pH of human blood plasma, SBF and 1.5SBF are listed in Table I.

Table I. Ion concentrations and pH value of human blood plasma, SBF and 1.5SBF.

Ion	Ion concentration (mmol/dm^3)		
	Human blood plasma	SBF	1.5SBF
Na^+	142.0	142.0	213
K^+	5.0	5.0	7.5
Mg^{2+}	1.5	1.5	2.25
Ca^{2+}	2.5	2.5	3.75
Cl^-	103.0	148.8	222
HCO_3^-	27.0	4.2	6.3
HPO_4^{2-}	1.0	1.0	1.5
SO_4^{2-}	0.5	0.5	0.75
pH	7.2 - 7.4	7.4	7.4

Surface characterization

Surface structural changes to the substrates following irradiation by the O_2 CM ion beams and 1.5SBF soaking were investigated with thin-film X-ray diffractometry (TF-XRD), X-ray photoelectron spectroscopy (XPS) and field emission scanning electron microscopy (FE-SEM). The contact angle of the substrates was measured by a contact angle meter.

RESULTS AND DISCUSSION

Figure 1 shows C1s XPS spectra of PE substrates unirradiated (a) and irradiated (b) by the simultaneous use of O_2 CM ion beams. The irradiated PE substrate gave a broad peak of $\equiv$C–OH group beside ethylene ($-(CH_2)_n-$) chain. This indicates that many chemical states besides $\equiv$C–OH group were formed on the PE surface by the irradiation, although the detailed surface structural change should be investigated in future. In addition, a small peak assigned to –COOH (around 290 eV) was newly observed after the irradiation. The presence of the COOH groups should be confirmed by other characterization techniques in future, since the present XPS peak

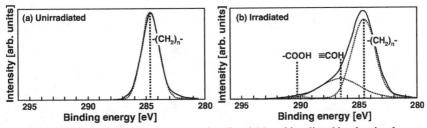

Figure 1. C1s XPS spectra of PE substrates unirradiated (a) and irradiated by the simultaneous use of O_2 CM ion beams (b).

for COOH was extremely small. The surface structural change produced by the simultaneous irradiation of O_2 CM ion beams can be interpreted as follows. The chemical bonds of PE were partially broken by O_2 monomer ions to form dangling bonds and, at the same time, a large number of oxygen atoms were supplied to the dangling bonds by O_2 cluster ions to form hydrophilic groups. The contact angle of the substrate was decreased from 100° to 10° by the irradiation. The formation of hydrophilic groups might be responsible for the low contact angle.

Table II shows surface compositions of silicone rubber substrates unirradiated and irradiated by the simultaneous use of O_2 CM ion beams. The atomic concentration of carbon was remarkably decreased from 50.56 to 28.51 %, whereas that of oxygen was remarkably increased from 24.20 to 48.98 % by the irradiation. This compositional change can be interpreted as follows. The silicone rubber has $Si-CH_3$ and $Si-O$ bonds in its structure. It is considered that the irradiated O_2 CM ions break the $Si-CH_3$ bond more preferentially than $Si-O$ bond to form dangling bond, since the binding energy of $Si-CH_3$ bond (84 kcal/mol) is lower than that of $Si-O$ bond (103 kcal/mol). The resultant dangling bond can react easily with oxygen atoms from O_2 CM ions and residual gas in the irradiation chamber to form new $Si-O$ bonds. As a result, carbon atoms are sputtered out and oxygen atoms are newly incorporated into the surface of silicone rubber.

Figure 2 shows Si2p XPS spectra of silicone rubber substrates unirradiated and irradiated by the O_2 CM ion beams. Unirradiated substrate gave a single peak at 102.3 eV which was attributed to $Si-O$ bond in silicone rubber. After the irradiation, two peaks were newly observed at 103.3 and 103.9 eV. They are attributed to SiO_m and SiO_n (n>m), respectively [8]. This indicates that silicon oxide (SiO_x) clusters were formed by the irradiation of O_2 CM ion beams. The mechanism of the formation of SiO_x clusters can be interpreted as follows. As described above, $Si-CH_3$ and $Si-O$ bonds in silicone rubber structure are broken to form $Si-$ dangling bonds by the irradiation. Thus formed dangling bonds can react easily with oxygen atoms from O_2 CM ions and/or surrounding oxygen atoms in $-(Si-O)_n-$ chain in silicone rubber, and hence

Table II. Surface compositions of silicone rubber substrates unirradiated and irradiated by the simultaneous use of O_2 CM ion beams.

Sample	Si / atomic%	C / atomic%	O / atomic%
Unirradiated	25.24	50.56	24.20
Irradiated	22.51	28.51	48.98

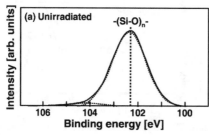 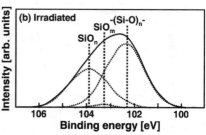

Figure 2. Si2p XPS spectra of silicone rubber substrates unirradiated (a) and irradiated by the simultaneous use of O_2 CM ion beams (b).

SiO$_x$ clusters are formed. The formation of SiO$_m$ should be further investigated, since the present XPS peak for SiO$_m$ was extremely small. The contact angle of the silicone rubber substrates was remarkably decreased from 112° to 10° by the irradiation. This suggests that some hydrophilic functional groups such as COOH and Si–OH were formed on the silicone rubber surface as well as PE surface.

Figure 3 shows TF-XRD patterns of PE and silicone rubber substrates unirradiated and irradiated by the simultaneous use of the O$_2$ CM ion beams, and then soaked in 1.5SBF for 7 days. Apatite peaks at 26 and 32° in 2θ were observed clearly in the TF-XRD measurement. This indicates that the irradiated PE and silicone substrates formed apatite on their surfaces, whereas unirradiated ones did not form it. It can be assumed that some functional groups such as COOH and Si–OH groups were formed on PE and silicone rubber substrates by the irradiation of O$_2$ CM ion beams in this study, and induced apatite nucleation in 1.5SBF.

Figure 4 shows FE-SEM photographs of PE and silicone rubber substrates unirradiated and irradiated with the O$_2$ CM ion beams and soaked in CaCl$_2$ solution, and then soaked in

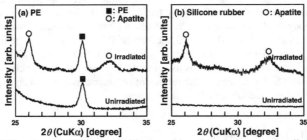

Figure 3. TF-XRD patterns of PE (a) and silicone rubber (b) substrates unirradiated and irradiated (b) by the simultaneous use of O$_2$ CM ion beams and soaked in CaCl$_2$ solution, and then soaked in 1.5SBF for 7 days.

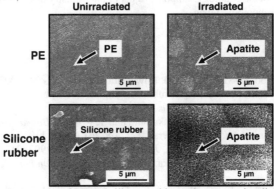

Figure 4. FE-SEM photographs of PE and silicone rubber substrates unirradiated and irradiated by the simultaneous use of O$_2$ CM ion beams and soaked in CaCl$_2$ solution, and then soaked in 1.5SBF for 7 days.

1.5SBF for 7 days. It can be confirmed that uniform apatite layer was formed on the irradiated substrates. The apatite took a scale-like morphology, which was often observed in SBF, although the mechanism for the unique morphology of the apatite is not reveled yet. These results indicate that the simultaneous irradiation of O_2 CM ion beams is a promising technique for inducing apatite-forming ability on polymer surface. The apatite was also formed on PE substrate without $CaCl_2$ treatment, but its amount was extremely small. In the case of silicone rubber, the apatite was not formed without $CaCl_2$ treatment. Therefore, it is considered that the $CaCl_2$ treatment is necessary for the uniform apatite formation.

CONCLUSIONS

PE and silicone rubber substrates were irradiated at an acceleration voltage of 7 kV and a dose of 1×10^{15} ions/cm^2 by the simultaneous use of oxygen cluster and monomer (O_2 CM) ion beams, and then soaked in $CaCl_2$ solution. The substrates were soaked for 7 days in a metastable calcium phosphate solution (1.5SBF) that had 1.5 times the ion concentrations of a normal simulated body fluid. The irradiated PE and silicone rubber substrates formed apatite on their surfaces, whereas unirradiated ones did not form it. This is attributed to the formation of functional groups effective for apatite nucleation, such as COOH and Si–OH groups, on the substrate surface by the simultaneous use of O_2 CM ion beams. We can conclude that apatite-forming ability can be induced on the surface of polymeric materials such as PE and silicone rubber by the simultaneous use of O_2 CM ion beams.

ACKNOWLEDGMENTS

This work was partially supported by the Ministry of Education, Culture, Sports, Science and Technology, Japan, Nippon Sheet Glass Foundation for Materials Science and Engineering, Japan and the Nanotechnology Support Project of the Ministry of Education, Culture, Sports, Science and Technology, Japan.

REFERENCES

1. T. Kokubo, H.-M. Kim and M. Kawashita, *Biomaterials* **24** (2003) 2161.
2. M. Tanahashi, T. Yao, T. Kokubo, M. Minoda, T. Miyamoto, T. Nakamura and T. Yamamuro, *J. Am. Ceram. Soc.* **77** (1994) 2805.
3. F. Balas, M. Kawashita, T. Nakamura and T. Kokubo, *Biomaterials* **27** (2006) 1704.
4. M. A. Duncan and D. H. Rouvray, *Sci. Am.* **261** (1989) 110.
5. I. Yamada and G. H. Takaoka, *Jpn. J. Appl. Phys.* **32** (1993) 2121.
6. H. Biederman and D. Slavinska, *Proc. 12th Int. Conf. Ion Imp. Tech. 98* (IEEE, 1999) pp. 815-819.
7. M. Kawashita, S. Itoh, R. Araki, K. Miyamoto and G. H. Takaoka, *J. Biomed. Mater. Res. Part A* in press.
8. T. G. Vladkova, I. L. Keranov, P. D. Dineff, S. Y. Youroukov, I. A. Avramova, N. Krasteva and G. P. Altankov, *Nucl. Instrum. Methods Phys. Res. B* **236** (2005) 552.

Mater. Res. Soc. Symp. Proc. Vol. 1020 © 2007 Materials Research Society   1020-GG07-17

Fabrication of Nanoscale Gold Clusters by Low Energy Ion Irradiation

Volha Abidzina[1], I. Tereshko[1], I. Elkin[2,3], V. Red'ko[1], S. Budak[4], C. Muntele[4], D. Walker[5], and D. ILA[4]

[1]Belarusian-Russian University, Prospect Mira 43, Mogilev, 212005, Belarus
[2]Research and Production Enterprise "KAMA VT" Plc., Karl Liebknecht Str. 3a, Mogilev, 212000, Belarus
[3]"NANTES - Systemy Nanotechnologii" Plc., Dolne Mlyny Str., 21, Boleslawiec, 59-700, Poland
[4]Center for Irradiation of Materials, Alabama A&M University, Normal, AL, 35762-1447
[5]Physics Dept., University of Alabama in Huntsville, Huntsville, AL, 35899

ABSTRACT

This work is focused on nanoscale gold particle formation by low-energy ion irradiation in glow-discharge plasma and research into particle growth by increasing the time of exposure.

SiO_2+Au films on SiO_2 substrates produced by ion beam assisted deposition (IBAD) were exposed to ion irradiation at 1.2 keV energy for 1-2 hours. Plasmon resonance emergence caused by nanoparticle formation was observed by optical absorption apectrometry (OAS).

INTRODUCTION

Inclusion of materials in glass has been used for hundreds of years to color glasses. Metallic colloids embedded in dielectrics produce colors associated with optical absorption at the surface plasmon resonance frequency [1-3]. Metallic ion implantation, thermal or laser annealing have been used to change linear and non-linear optical properties near the surface of silica glass [4-10]. Metal nanoparticles have mainly been studied because their unique optical properties cause a broad absorption band in the visible region of the electromagnetic spectrum [11].

Optical properties of glass with metallic particles, as a rule, are characterized by optical absorption or reflection [12, 13]. Intensity and position of a selective region maximum of metallic nanoparicles are defined by effects of plasmon resonance and depend on concentration and particles size. In case of spherical shaped particles and their low concentration, spectral position of ranges is successfully predicted by Mie theory [14]. In particular, the theory allows to define quantitatively an average size of particles in case of their uniform volume distribution and narrow function of particles size distribution.

Spectra of metallic nanoparticles can also be analyzed using effective media theory [14], i.e. increase in metallic fraction in the sample leads to red shift of plasmon resonance spectrum.

EXPERIMENTAL METHODS

SiO_2+Au films on SiO_2 substrates were produced by IBAD. The Amersil silica rod and gold (from Scientific Instrument Services) were placed in the crucible of the deposition system. The

SiO$_2$ substrates (suprasil) were placed on the substrate holder, ten inches above the silica rod. The co-deposition of gold and silica was conducted at 6·10^{-6} Torr. The evaporation rate of gold and silica was calibrated in advance and subsequently checked by RBS. The thicknesses of the films and Au concentration were measured by RBS. The RBS analysis of a typical sample is shown in Figure 1. Samples prepared for the investigation had 34±2 nm of thickness and 1.4±0.4 of at% gold concentration.

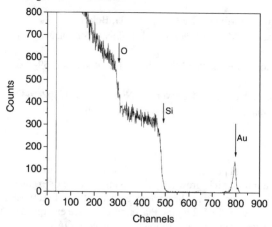

Figure 1. RBS spectrum of a typical sample made for the investigation.

To study nanocluster formation, samples were exposed to low-energy ion irradiation in glow-discharge plasma of residual gases. The ion energy depended on the voltage in the plasma generator and did not exceed 1.2 keV while the current in the plasma generator was maintained at 70 mA. The temperature in the chamber was controlled during the irradiation process and did not exceed 323 K. The irradiation time varied from 60 to 120 minutes.

The formation of Au nanoclusters after low-energy ion irradiation was studied using optical absorption spectrometry (OAS).

SRIM program was used to simulate plasma influence onto thin films. Since residual gas of atmosphere was used during irradiation, SRIM calculation was made for nitrogen ions. For this purpose a "trim.dat" file containing various energies of incident ions and incident angles was created [15]. Energy of incident ions was randomized and did not exceed 3 keV. All ions were supposed to direct to the surface and interact with it. Calculation for nitrogen ion beam with 3 keV energy was also made for comparing processes taken place after plasma treatment and ion bombardment.

RESULTS AND DISCUSSION

It was previously shown [16, 17] using Mie theory and index n_0 for silica glass that surface plasmon resonance for gold nanoparticles is supposed to be at 535 nm.

Figure 2 represents the optical absorption spectra of unirradiated sample and samples that were irradiated for 1 and 2 hours at 1.2 keV. Broad selective absorption spectra observed in the visible region are caused by nanoparticles formation and their plasmon resonance emergence (spectra 2 and 3).

Increasing the irradiation time, we observed a red shift in wavelength as shown in our previous publication [16, 17]. In addition, we observed that the full width at half maximum (FWHM) of the absorption band increased as we were increasing the irradiation time. Broad absorption band indicates to nanoscale particles formation on the surface of the samples. Increasing the time of exposure, the red shift of plasmon resonance is observed, which according to Mie theory corresponds to particles growth.

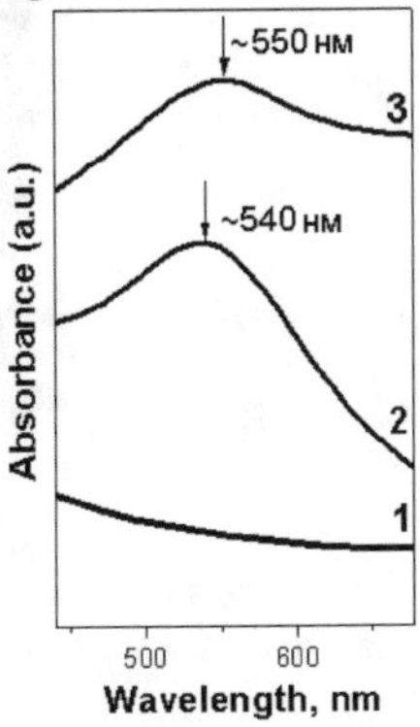

Figure 2. Optical absorption spectra: 1 – unirradiated sample, 2 – irradiated sample for 1 hour at 1.2 kV, 3 – irradiated sample for 2 hours at 1.2 kV.

Figure 3 shows nitrogen ions distribution in SiO$_2$+Au thin film. In case of plasma treatment (figure 3a), ions penetrate not more than 15 Å in depth. At nitrogen ions bombardment (directed mono beam), ions penetrate 400 Å in depth (figure 3b). Thus, a very thin near-surface region is exposed to plasma treatment. The ion range does not exceed the film thickness, therefore the red shift of plasmon resonance (figure 2, spectra 2 and 3) cannot be due to optical properties changes of the substrate.

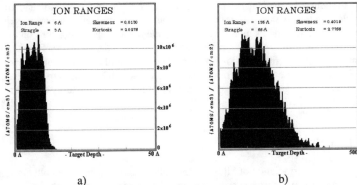

a)   b)

Figure 3. Distribution of nitrogen ions in SiO$_2$+Au layer: a – plasma ion treatment (ion energy has random value and does not exceed 3 keV), b – nitrogen ion beam at 3 keV energy.

Accelerated ions embedded into a solid lose their energy at electronic and nuclear collisions. Nuclear energy loss may lead to target atoms displacement, electronic energy loss may lead to the initiation of electron transition that leads to excitation and ionization of atoms [18]. Figure 4 represents nuclear and electronic energy losses for plasma and ion beam.

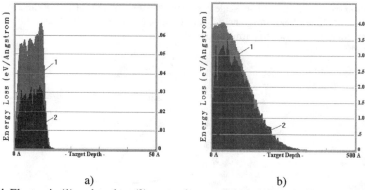

a)   b)

Figure 4. Electronic (1) and nuclear (2) energy losses of nitrogen ions in SiO_2+Au layer: a – plasma ion treatment (ion energy has random value and does not exceed 3 keV), b – nitrogen ion beam at 3 keV energy.

Comparing energy losses shown in figure 4b, it is seen that in case of ion beam, electronic and nuclear energy losses take place almost equally. However, in case of plasma treatment, electronic energy loss is two times more than nuclear one, which leads to active ionization of thin film atoms. Au atoms may form chemical bonds with emerging radicals and silica ions or take part in oxidation processes. However, due to the difference in Gibbs free energy Au-Au bonds are mainly formed, which results in cluster enlargement. Free energy of SiO_2 formation is far less than for metallic Au and its oxide, therefore oxygen bonding with Si predominates [19]. If even short-term bonds of Au and O are formed, they tend to dissociate to form Si-O and Au-Au bonds to reduce the total energy of the system. Assuming that nucleation and growth of metal nanoparticles are the result of Au^o joining, then the growth of metal nanoparticles depends on local Au concentration and its mobility. Nanoparticle formation also depends on time and the fluence of irradiation and sputtering process. Sputtering of thin film atoms from the surface may also promote Au atoms concentration on the surface. It should be also taken into consideration the generation of nonlinear oscillations in nanocrystals which become active centers. These oscillations are due to glow-discharge molecules that have 90% of energy concentrated on vibrating degree of freedom [20].

CONCLUSIONS

Low-energy ion irradiation of SiO_2+Au thin films leads to nanoscale gold clusters formation as observed by optical absorption spectrometry. The longer exposure to the glow-discharge provides the red shift in the absorption band as well as increase in the FWHM of the absorption band. The broad absorption band indicates the nanoscale particles formation on the surface of the samples.

SRIM calculation shows active ionization of thin film atoms by plasma treatment, which may lead to nanocluster formation. The growth of nanoparticles depends on local Au concentration, its mobility, time and the fluence of irradiation and sputtering process.

ACKNOWLEDGEMENTS

Research has been sponsored by the Center for Irradiation of Materials, Alabama A&M University and by the AAMURI Center for Advanced Propulsion Materials under the contract number NNM06AA12A from NASA, and by National Science Foundation under Grant No. EPS-0447675.

REFERENCES

1. G.W. Arnold, *J. Appl. Phys.* **46**, 4466 (1975).
2. G. Fuchs, G. Abouchacra, M. Treilleux, P. Thevenard, J. Serughetti, *Nucl. Instr. and Meth.* **B32**, 100 (1988).
3. G. Abouchacra, G. Chassagne, J. Serughetti, *Radiat. Eff.* **64**, 189 (1982).
4. G.W. Arnold, J.A. Bordes, *J. Appl. Phys.* **48**, 1488 (1977).
5. D. Ila, Z. Wu, C.C. Smith, D.B. Poker, D.K. Hensley, C. Klatt, S. Kalbitzer, *Nucl. Instr. and Meth.* **B 127/128**, 570 (1997).
6. R.H. Magruder III, R.A. Zuhr, D.H. Osborne, Jr., *Nucl. Instr. and Meth.* **B 99** (1995) 590.
7. Y. Takeda, T. Hioki, T. Motohiro, S. Noda, T. Kurauchi, Nucl. *Instr. and Meth.* **B 91**, 515 (1994).
8. D. Ila, Z. Wu, R.L. Zimmerman, S. Sarkisov, C.C. Smith, D.B. Poker, D.K. Hensley, *Mat. Res. Soc. Symp. Proc.* **457**, 143 (1997).
9. C.W. White, D.S. Zhou, J.D. Budai, R.A. Zuhr, R.H. Magruder, D.H. Osborne, *Mat. Res. Soc. Symp. Proc.* **316**, 449 (1994).
10. K. Fukumi, A. Chayahara, M. Adachi, K. Kadono, T. Sakaguchi, M. Miya, Y. Horino, N. Kitamura, J. Hayakawa, H. Yamashita, K. Fujii, M. Satou, *Mat. Res. Soc. Symp. Proc.* **235**, 389 (1992).
11. C.F Bohren, D.R Huffman, *Absorption and Scattering of Light by Small Particles*, (Wiley: New York, 1983).
12. P.D. Townsend, P.J. Chander, L. Zhang, *Optical effects of ion implantation*, (Cambridge: Cambridge University Press, 1994) p.165.
13. A.L. Stepanov, *Optics and Spectroscopy*, **89** (3), 444 (2000).
14. U. Kreibig, M. Vollmer, *Optical properties of metal clusters*. (Berlin: Springer-Verlag, 1995) p. 533.
15. F. Ziegler, J.P. Biersack, U. Littmark, *The Stopping and Range of Ions in Solids*, (Pergamon Press, New York, 1985).
16. V. Abidzina, I. Tereshko, I. Elkin, R.L. Zimmerman, S. Budak, B. Zheng, C. Muntele, D. Ila, *Mat. Res. Soc. Symp. Proc.* **929**, 191-195 (2006).

17. V. Abidzina, I. Tereshko, I. Elkin, S. Budak, C. Muntele, D. Ila, *Nucl. Instr. and Meth. B* (2007) (in press).
18. W. Eckstein, *Computer Simulation of Ion-Solid Interactions*, (Springer-Verlag, 1991).
19. F.Gonella, P.Mazzoldi, *Handbook of Nanostructured Materials and Nanotechnology*, (San Diego: Academic Press, 2000) pp.65-74.
20. A.V. Eletskiy, *Glow Discharge*, (Znanie: M., 1981).

Mater. Res. Soc. Symp. Proc. Vol. 1020 © 2007 Materials Research Society   1020-GG07-18

Fluence Dependence of Thermoelectric Properties Induced by Ion Bombardment of Zn4Sb3 and CeFe2Co2Sb12 Thin Films

C. C. Smith[1], S. Budak[2], S. Guner[2], C. Muntele[2], R. A. Minamisawa[2], R. L. Zimmerman[2], and D. ILA[2]

[1]MSFC, NASA, MSFC, Huntsville, AL, 35812

[2]Center for Irradiation of Materials, Alabama A&M University, 4900 Meridian Street, PO Box 1447, Normal, AL, 35762

Abstract

Thermoelectric power generation is a promising technology for increasing the efficiency of electrical and optical electrical devices. We prepared samples by Electron Beam evaporated Zn_4Sb_3 and $CeFe_2Co_2Sb_{12}$ thin films on silicon dioxide (silica) substrates. The materials were co-evaporated and then were prepared for gold over-coating. Following electron deposition we performed post ion bombardment at a constant energy of 5 MeV while varying fluence from 1×10^{12}, 1×10^{13}, 1×10^{14}, 1×10^{15} ions/cm^2, respectfully. The production of nano-clusters generated by the MeV Si ions bombardment modifies the electrical and phonon interactions in the materials. Also, we will report on the fluence dependence of the figure of merit, Seebeck coefficient, thermal conductivity and electrical conductivity. In addition, Rutherford backscattering spectrometry (RBS) was used to analyze the elemental composition and the thickness of the deposited material.

*Corresponding author: D. ILA; Tel.: 256-372-5866; Fax: 256-372-5868; Email: ila@cim.aamu.edu

1. INTRODUCTION

Processing thermo electric materials into thin films allows for unique devices that would be difficult to process using bulk materials [1]. The efficiency of the thermoelectric devices is determined by the figure of merit ZT [2]. The figure of merit is $ZT = S^2\sigma T/\kappa$, where **S** is the Seebeck coefficient, σ is the electrical conductivity, **T** is the absolute temperature, and κ is the thermal conductivity [3,4]. ZT can be increased by increasing **S,** increasing σ, or decreasing κ. Efficient thermoelelctric devices have a high electrical conductivity and a low thermal conductivity [5].

Our investigation focuses on the development of nano structures and thin films of previously investigated bulk materials. Our goal is to process films with the following properties: Low thermal conductivity, high electrical conductivity and a high figure of merit.

2. EXPERIMENTAL

Zn_4Sb_3 and $CeFe_2Co_2Sb_{12}$ thin films were prepared using an ion beam assisted deposition (IBAD) system. In the preparation process of Zn_4Sb_3, Zn and Sb were placed in the two separate Telemark Electron gun evaporators and were co-evaporated to form the thin film of Zn_4Sb_3. In the preparation process of $CeFe_2Co_2Sb_{12}$, Ce, Fe and Co were placed in one of the two separate Telemark Electron gun evaporators while Sb was placed in the other electron gun evaporator and they were co-evaporated to form the thin film of $CeFe_2Co_2Sb_{12}$. The base pressure for both the depositions was $6.0x\ 10^{-6}$ torr. The thickness of the films was controlled by an INFICON deposition monitor. The film geometries used in this study are shown in Fig.1. Fig.1a shows the geometry of Zn_4Sb_3 thin film from the cross-section, while Fig.1b shows the geometry of the $CeFe_2Co_2Sb_{12}$thin film. The geometries in Fig.1 show Au contacts on the top and the bottom of the films. These two Au contacts were used in the Seebeck coefficient measurement. The electrical conductivity was measured by the Van der Pauw system and the thermal conductivity was measured by the 3ω technique. The electrical conductivity, the thermal conductivity and the Seebeck measurements were performed at a temperature of 22 ^{0}C. Detailed information about 3ω technique may be found in Refs. [6-8]. The 5 MeV Si ions bombardments were accomplished with the Pelletron ion beam accelerator at the Alabama A&M University Center for Irradiation of Materials (AAMU-CIM).

SRIM monte carlo simulation software was used to choose to the bombarding energies of the Si ions. The fluences used for the bombardments were $1x10^{12}$, $1x10^{13}$, $1x10^{14}$, and $1x10^{15}$ ions/cm^2. Rutherford backscattering spectrometry (RBS) was performed using 2.1 MeV He$^+$ ions with the particle detector placed at 170 degrees from the incident beam, to monitor the film thickness and stoichiometry before and after 5 MeV Si ion bombardments [9,10].

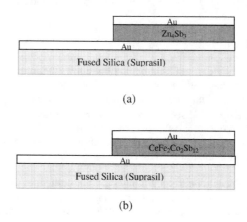

(a)

(b)

Fig. 1. Geometry of sample from the cross-section.

3. RESULTS AND DISCUSSION

Fig. 2a. shows the RBS spectrum of Zn_4Sb_3 thin film on Glassy Polymeric Carbon (GPC) substrate when the sample is at the normal angle. RUMP simulation [11] gives information about the amount of the elements in the deposited samples and the thickness of the deposited films. According to the RUMP simulation, the thickness for Zn_4Sb_3 thin film was found as 383 nm at the stoichiometry of 4:3 for Zn and Sb respectively. Fig. 2b. shows the RBS spectrum of the $CeFe_2Co_2Sb_{12}$ thin film on a Glassy Polymeric Carbon (GPC) substrate when the sample is at the normal angle. The RUMP simulation, gave the thickness of 384 nm for $CeFe_2Co_2Sb_{12}$ thin film. We have successfully deposited films and tested and analyzed the thermoelectric and electrical properties of Zn_4Sb_3 and $CeFe_2Co_2Sb_{12}$ thin films. Fig. 3 shows the thermoelectric properties of Zn_4Sb_3 and ZrNiSn thin films. All the parameters in Fig. 3 have been plotted on the log-log scale. The Zn_4Sb_3 thin films exhibit enhanced thermoelectric properties and electrical properties as a function of ion fluence. The $CeFe_2Co_2Sb_{12}$ thin film showed diminished properties for the thermoelectric and electrical properties. The Seebeck, electrical conductivity and the thermal conductivity all exhibited expected changes as a function of the ion fluence when the suitable fluence of ion bombardment was choosen. For the $CeFe_2Co_2Sb_{12}$ thin film, the Seebeck, thermal conductivity and the figure of merit showed no improvement as function of fluence after the fluence of 1×10^{14} ions/cm^2.

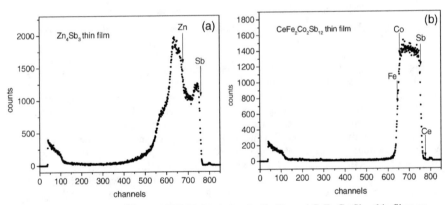

Fig. 2. He RBS spectra and RUMP simulation for Zn_4Sb_3 and $CeFe_2Co_2Sb_{12}$ thin films on GPC substrate.

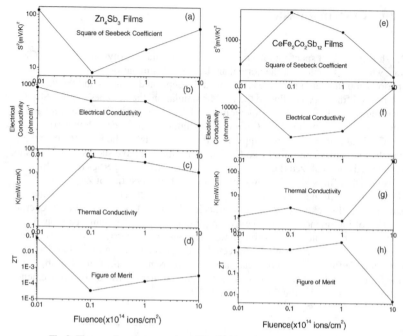

Fig.3. Thermoelectric properties of Zn_4Sb_3 and $CeFe_2Co_2Sb_{12}$ thin films.

In Fig. 3a, the square of the Seebeck coefficient for Zn_4Sb_3 decreases until the fluence of 1×10^{13} ions/cm^2. After the fluence of 1×10^{13} ions/cm^2, it increases as the fluence increases. As seen from Fig.3e, for the $CeFe_2Co_2Sb_{12}$ thin films the square of the Seebeck coefficient increases with increasing fluence until the fluence of 1×10^{13} ions/cm^2. After the fluence of 1×10^{13} ions/cm^2, it decreases as the fluence increases. In Fig. 3b the electrical conductivity for the Zn_4Sb_3 thin film decreases with increasing fluence. The electrical conductivity of the $CeFe_2Co_2Sb_{12}$ decreases as a function of ion fluence until the fluence of 1×10^{13} ions/cm^2. Figure 3c for the Zn_4Sb_3 thin film, the thermal conductivity value increases initially until the fluence of 1×10^{13} ions/cm^2 and then as the fluence increases the thermal conductivity decreases. As seen from Fig. 3g, the thermal conductivity for the $CeFe_2Co_2Sb_{12}$ thin film increases until the fluence of 1×10^{13} ions/cm^2 and after the fluence of 1×10^{13} ions/cm^2 it decreases and starts to increase after the fluence of 1×10^{14} ions/cm^2. For the figure of merit for Zn_4Sb_3 thin film in Fig. 3d, the data show a small amount of change as a function of fluence. The bulk β- Zn_4Sb_3 has maximum dimensionless figure of merit of 0.2 in the room temperature [12]. We obtained maximum figure of merit as 0.1 at the fluence of 1×10^{12} ions/cm^2 for our thin film structure in the room temperature. As seen from Fig. 3h, the figure of merit for the $CeFe_2Co_2Sb_{12}$ thin film 1.64 at the fluence of 1×10^{12} ions/cm^2, 1.37 at the fluence of 1×10^{13} ions/cm^2 and exhibits maximum value as 3.08 at the fluence of 1×10^{14} ions/cm^2. When we compare the

ZT of bulk or thin film skutterites without any ion bombardment, the maximum value was observed as ~0.5 in the room temperature [12].

4. CONCLUSION

Si ion bombardment of $CeFe_2Co_2Sb_{12}$ and Zn_4Sb_3 thin films produced changes in the thermoelectric properties of thin films. The positively expected results for dimensionless figure of merit have been obtained mainly for the thin film of $CeFe_2Co_2Sb_{12}$ with respect to bulk form of same material. The ZT reached a highest value of 3.08, which is 3 times more than well promising thermoelectric materials at the fluence of 1×10^{14} ions/cm^2 in the room temperature. Further investigations are being implemented to further improvement of thermoelectric device fabrication.

Acknowledgement

Research sponsored by the Center for Irradiation of Materials, Alabama A&M University and by the AAMURI Center for Advanced Propulsion Materials under the contract number NNM06AA12A from NASA, and by National Science Foundation under Grant No. EPS-0447675.

References

1. T.M. Tritt, ed., "Recent Trends in Thermoelectrics in Semiconductors and Semimetals", 71 (2001).
2. G. D. Mahan, Solid State Physics, 51, 81 (1998).
3. L. D. Hicks and M. S. Dresselhaus, Physics Review B47, 12727 (1993).
4. B C. -K. Huang, J. R. Lim, J. Herman, M. A. Ryan, J. -P. Fleural, N. V. Myung, Electrochemical Acta 50, 4371. (2005)
5. Z. Xiao, R.L. Zimmerman, L.R. Holland, B. Zheng, C. I. Muntele, D. Ila, Nuc. Instr. and Meth. B 242,201–204 (2006).
6. L. R. Holland, R. C. Smith, J. Apl. Phys. 37, 4528(1966).
7. D. G. Cahill, M. Katiyar, J. R. Abelson, Phys. Rev.B 50, 6077 (1994).
8. T. B. Tasciuc, A.R. Kumar, G. Chen, Rev. Sci. Instrum. 72, 2139 (2001).
9. J. F. Ziegler, J. P. Biersack, U. Littmark, The Stopping Range of Ions in solids, Pergamon Press, New York, 1985.
10. W. K. Chu, J. W. Mayer, M. -A. Nicolet, Backscattering Spectrometry, Academic Press, New York, 1978.
11. L. R. Doolittle, M. O. Thompson, RUMP, Computer Graphics Service, 2002.
12. L.T. Zang, M. Tsutsui, K. Ito, M. Ymaguchi, Thin Solid Films 443, 84 (2003).

Mater. Res. Soc. Symp. Proc. Vol. 1020 © 2007 Materials Research Society   1020-GG07-19

Effect of Implantation Dose and Annealing Time on the Formation of Si Nanocrystals Embedded in Thermal Oxide

Cong Qian[1], Zheng-xuan Zhang[2], Feng Zhang[2], and Cheng-lu Lin[2]

[1]National Key Laboratory of Functional Materials for Informatics, Shanghai Institute of Microsystem and and Information Technology, Chinese Academy of Sciences, Rm 514, No.6 Building, No.865 Changning Rd, Changning District, Shanghai City, P.R. China, Shanghai, 200050, China, People's Republic of

[2]Shanghai Institute of Microsystem and Information Technology, Chinese Academy of Sciences, Shanghai, 200050, China, People's Republic of

ABSTRACT

Photoluminescence (PL) and X-ray Photoelectron Spectroscopy (XPS) are employed to study the Si nanocrystals formed in the thermal oxide by Si^+ implantation. PL results estimate the size of nanocrystals and the concentration of P_b centers in the $Si-SiO_2$ nanocrystal-matrix interfaces. It is shown that the size of Si nanocrystals increase with implantation dose. Increasing the dose from 1×10^{16} to 1×10^{17} Si^+/cm^2 shifts the size of nanocrystals from ~2 nm to ~3.5nm, while prolonging the annealing time from 1h to 2h has no effect on the position of PL peak. P and Ar implantations into the SiO_2 films are also investigated to suggested that the PL peak is due to implant induced chemical changes rather than implant induced damage. XPS analysis shows that the concentration of Si nanocrystals increases with Si implantation dose. Research on the annealing dependence of the forming of Si nanocrystals suggests that $1000°C$ annealing produces larger amount of Si nanocrystals than $1100°C$ annealing.

INTRODUCTION

Si nanocrystals(or nanoclusters) embedded in SiO_2 films have attracted much attention recently[1], which exhibit new quantum phenomena and have potential applications in optoelectronic devices, single electron memory devices and other single electron devices. A promising techniques for producing Si nanocrystals is Si ion implantation into SiO_2 films grown by thermal oxidation of a Si wafer [1].

The PL spectrum arising from the Si nanocrystals has been an important method to investigate these structures. The wavelength of the peak PL emission has been shown to be correlated with the diameter of the nanocrystals, redshifting from ~700 to ~870 nm as the average nanocrystals diameter increases from 3 to 7 nm [2]. The redshifting can be explained as resulting from quantum confinement effects [3]. It was reported [2] that the PL intensity inversely correlates with the amount of P_b paramagnetic centers at the $Si-SiO_2$ nanocrystal-matrix interfaces, which can be measured by electron spin resonance(ESR).

EXPERIMENT

Sample preparation

The samples studied were 380-nm-thick thermal oxides grown on 10-25 ohm·cm p-type Si (100) substrates at 1000℃. The energies of implanted Si are 50, 75, 100, 125 and 150keV. According to the TRIM simulation, the Si peak concentration are about 90, 135, 180, 225 and 270 nm below the surface of the thermal oxide, corresponding to the five implant energies respectively. The implantation doses are 1×10^{16}, 5×10^{16} and 1×10^{17} Si$^+$/cm^2. After implantation, the samples are annealed in Ar at 1000☐ for either 60 min or 120 min. An Unimplanted sample, a 5×10^{16} P/cm^2 implanted sample and a 5×10^{16} Ar/cm^2 implanted sample were also employed, which were annealed in Ar at 1000☐ for 120 min..

PL measurement

PL measurements were carried out using a He-Cd laser operating at ~ 10mW as the excitation source. The measurements were done at room temperature. The end of a fiber optic cable was placed near the sample (1cm×1cm) so as to collect light emitted normal to the sample. Luminescent light was separated from scattered laser light by passing it through a 325-nm high-pass filter. A GaAs detector was employed. The system was sensitive to light in the range of 325 to 1000 nm.

PL RESULT AND DISCUSSION

Implantation dose dependence

Figure 1(a) shows the room temperature PL spectrum from samples implanted with 1×10^{16}, 5×10^{16} and 1×10^{17} Si$^+$/cm^2 and annealed Ar for 120 min at 1000℃. Main PL peaks (600-800 nm) are attributed to the emission from silicon nanocrystals [4-6]. Note that increasing the implant dose from 1×10^{16} to 5×10^{16} and 1×10^{17} Si$^+$/cm^2 shifts the PL peak from ~ 600 to ~ 650 and ~725 nm, reflecting the size of nanocrystal increases from ~2 nm to ~2.5nm and then ~3.5nm, as a result of additional agglomeration of implanted Si.

5×10^{16}/cm^2 P ion implantation results in a small PL peak (Figure 1 (b)), suggesting the similar nanostructure has been formed by P implantation into SiO$_2$ films. The PL peak locates at ~575 nm, indicating the nanocrystals produced by P implantation is smaller in size than those produced by Si implantation. Reference [7] investigates the Si, Al, P and Ar implantation into the thermal oxides.

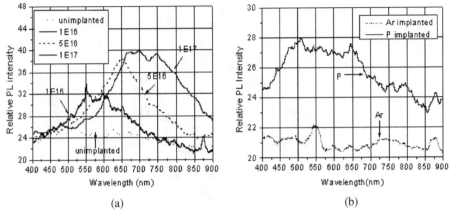

(a)   (b)

Figure 1. PL intensity as a function of wavelength for oxides implanted with
(a) various doses of Si ions, then annealed in Ar for 120 min at 1000°C
(b) $5\times10^{16}/cm^2$ P or Ar ions, then annealed in Ar for 120 min at 1000▯

No PL peak was observed in either unimplanted oxide (Figure 1 (a)) or Ar-implanted sample (Figure 1 (b)). It is suggested that the PL peak is due to implant induced chemical changes rather than implant induced damage as Si, P and Ar implants produce similar amount of damage to the oxide according to SRIM calculations [8], while   unlike Si and P, Ar implant will not cause any chemical change to the oxide.

Implantation energy dependence

Figure 2 gives PL intensity as a function of wavelength for oxides implanted with various energies of $5\times10^{16}/cm^2$ Si$^+$. As seen in Figure 2, the 125keV implant has the largest PL intensity. Because PL intensity inversely correlates with the amount of P$_b$ centers at the Si-SiO$_2$ interfaces, the 125keV Si-implanted oxide should contain the smallest amount of P$_b$ centers. Notice that all the five PL peaks locate at ~650nm, irrelevant to the implant energy.

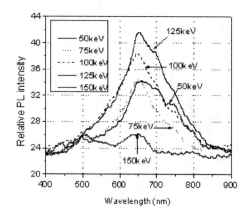

Figure 2. PL intensity as a function of wavelength for oxides implanted with 5×10^{16}/cm^2 Si ions, at various energies (annealed in Ar for 120 min at 1000℃)

Annealing time dependence

Figure 3 gives the PL results of samples implanted with 5×10^{16}/cm^2 Si ions, at an energy of 100 keV, then annealed in Ar at 1000□ for 1h and 2h, respectively. It is shown that the position of PL peak is independent of the annealing time. Since the size of the Si nanoclusters evidently grows as the annealing time increases [9], the independence of the PL peak position on the annealing time excludes the possibility that the luminescence is simply due to the direct recombination between electrons and holes confined in the inside of Si [10]. The PL intensity of 2h annealing sample is about 5% larger than that of 1h annealing sample.

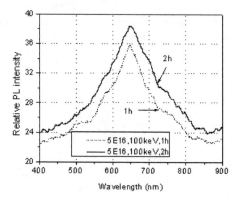

Figure 3. PL intensity as a function of wavelength for oxides implanted with 5×10^{16}/cm^2 Si ions, at an energy of 100 keV, then annealed in Ar for 60 min and 120 min at 1000□

XPS RESULT AND DISCUSSION

XPS analysis was performed using a XSAM800 spectrometer with monochromatic Mg Kα(1253.6eV) x-ray radiation. The energy scale of the XPS spectra was calibrated with the binding energy of the C 1s peak The Si 2p XPS peaks were analysed by means of a Gaussian peak fitting computer program. From deconvolution of the Si 2p peaks, the concentrations of various components corresponding to the different Si oxidation states Si^{n+} (n=0, 1, 2, 3 and 4) in the SiO_x films were determined.

The XPS results clearly show the evolution of various chemical structures and the formation of Si nanocrystals in the oxide of our samples. It would be useful to examine quantitatively the changes of concentration of the oxidation states with the Si implantation dose. The relative concentration (in percentage) of each oxidation state is obtained by calculating the ratio I_{Si}^{n+} / I_{total} (n=0,1,2,3,4), where I_{Si}^{n+} is the peak area of the oxidation state Si^{n+} and I_{total} is the total area ($= \sum_{i=0}^{4} I_{Si}^{i+}$) of the Si 2p peaks.

Table 1 shows the changes in the concentrations of the various oxidation states with the Si implantation dose for fixed implantation energy and annealing conditions. It can be seen in the table that the concentration of Si^0 increases with Si implantation dose, suggesting more Si nanocrystals have formed [1]. We can see that 1000□ annealing produces larger amount of Si nanocrystals than 1100□ annealing. Annealing at a higher temperature (>800□) leads to a decrease in the concentration of Si^0, suggesting that effect of thermal oxidation is more significant than thermal decomposition at higher temperature[1].

Table 1. The concentration of the five Si oxidation states Si^{n+} (n=0,1,2,3,4) with Si implantation dose

(1) 1000□ Anealing

	5E15	1E16	5E16
Si0	0.0227	0.0502	0.0511
Si1+	0.0563	0.0778	0.0833
Si2+	0.0641	0.0639	0.0667
Si3+	0.2659	0.2209	0.2316
Si4+	0.5910	0.5872	0.5673

(2) 1100□ Anealing

	1E16	5E16	1E17
Si0	0.0147	0.0202	0.0294
Si1+	0.0415	0.0526	0.0599

Si2+	0.0809	0.0728	0.0965
Si3+	0.3952	0.3563	0.3406
Si4+	0.4677	0.4980	0.4803

SUMMARY

In conclusion, PL analysis demonstrates the effect of Si implantation dose, energy and the annealing time on the formation and size of Si nanocrystals. The PL peak redshifts from ~600nm to ~725nm when the implantation dose increases from 1×10^{16} to 1×10^{17} Si^+/cm^2, reflecting the growth of nanocrystals. Prolonging annealing time from 1h to 2h results in the increase of PL intensity without shifting of PL peaks. The energy of the implanted Si, which effect the P_b centers in the Si-SiO$_2$ nanocrystal-matrix interfaces, has no impact on the growth of nanocrystals. P and Ar implantation into SiO$_2$ films are also discussed to verify that PL spectrum comes from chemical changes rather than implant damages. XPS analysis shows that 1000□ annealing produces larger amount of Si nanocrystals than 1100□ annealing.

REFERENCES

1. Y. Liu, T. P. chen, Y.Q. Fu, M.S. Tse, J.H. Hsieh, P.F. Ho and Y.C. Liu, J. Phys. D: Appl. Phys., 36, 97 (2003)

2. B. Garrido Fernandez, M. Lopez, C. Garcia, A. Perez-Rodriguez, J.R. Morante, C. Bonafos, M. Carrada and A. Claverie, J. Appl. Phys., 91(2), 798 (2002)

3. B. J. Mrstik and H. L. Hughes, IEEE Trans. Nucl. Sci., 50(6), 1947 (2003)

4. V. Ioannou-Sougleridis, B. Kamenev, D.N. Kouvatsos and A.G. Nassiopoulou., Mater. Sci. and Eng., B101, 324 (2003)

5. P. Photopoulos, A.G. Nassiopoulou, D.N. Kouvatsos and A. Travlos, Appl, Phys. Lett., 76(24), 3588 (2000)

6. P. Photopoulos and A.G. Nassiopoulou, Appl. Phys. Lett., 77(12), 1816 (2000)

7. B. J. Mrstik and H. L. Hughes IEEE Trans. Nucl. Sci., 47(6) , 2189 (2000)

8. J. F. Ziegler and J. P. Biersack, SRIM 2003 Program

9. C. J. Nicklaw, M. P. Pagey, S. T. Pantelides, D. M. Fleetwood, R. D. Schrimpf, K. F. Galloway, J. E. Wittig, B. M. Howard, E. Taw, W. H. McNeil, and J. F. Conley, IEEE Trans. Nucl. Sci., 47(6), 1947 (2269)

10. Tsutomu Shimizu-Iwayama, Norihiro Kurumado, David E. Hole and Peter D. Townsend, J. Appl. Phys., 83 , p.6018 (1998)

Mater. Res. Soc. Symp. Proc. Vol. 1020 © 2007 Materials Research Society   1020-GG07-21

Properties of Nano-Layers of Nanoclusters of Ag in SiO2 Host Produced by MeV Si Ion Bombardment

C. C. Smith[1], S. Budak[2], S. Guner[2], C. Muntele[2], R. A. Minamisawa[2], R. L. Zimmerman[2], and D. ILA[2]

[1]MSFC, NASA, MSFC, Huntsville, AL, 35812
[2]Center for Irradiation of Materials, Alabama A&M University, 4900 Meridian Street, PO Box 1447, Normal, AL, 35762

Abstract

We prepared 50 periodic nano-layers of $SiO_2/Ag_xSiO_{2(1-x)}$. The deposited multi-layer films have a periodic structure consisting of alternating layers where each layer is between 1-10 nm thick. The purpose of this research is to generate nanolayers of nanocrystals of Ag with SiO_2 as host and as buffer layer using a combination of co-deposition and MeV ion bombardment taking advantage of the electronics energy deposited in the MeV ion track due to ionization in order to nucleate nanoclusters. Our previous work showed that these nanoclusters have crystallinity similar to the bulk material. Nanocrystals of Ag in silica produce an optical absorption band at about 420 nm. Due to the interaction of nanocrystals between sequential nanolayers there is widening of the absorption band. The electrical and thermal properties of the layered structures were studied before and after 5 MeV Si ions bombardment at various fluences to form nanocrystals in layers of SiO_2 containing few percent of Ag. Rutherford Backscattering Spectrometry (RBS) was used to monitor the stoichiometry before and after MeV bombardments.

Keywords: Ion bombardment, thermoelectric properties, multi-nanolayers, SiO2/Ag+SiO2, Rutherford backscattering spectrometry, Van der Pauw method, 3ω method, Seebeck coefficient, Figure of merit.

*Corresponding author:
D. ILA; Tel.: 256-372-5866; Fax: 256-372-5868; Email: ila@cim.aamu.edu

1. INTRODUCTION

Metal and semiconductor nanoclusters embedded in transparent matrices exhibit linear and nonlinear optical properties that are of interest to the field of opto-electronics. By using the same technique it is feasible to produce these clusters for converting heat into electrical power [1]. The efficiency of the thermoelectric devices is determined by the figure of merit ZT [2]. The figure of merit is $ZT = S^2 \sigma T / \kappa$, where S is the Seebeck coefficient, σ is the electrical conductivity, T is the absolute temperature, and κ is the thermal conductivity [3-4]. Z T can be increased by increasing **S**, by increasing σ, or by decreasing κ. Efficient thermoelelctric devices have a high electrical conductivity and a low thermal conductivity [5].

2. EXPERIMENTAL

We have grown $SiO_2/Ag_xSiO_{2(1-x)}$ nano-layers films on silica substrates with the Ion Beam Assisted Deposition (IBAD) system. The multilayer films were sequentially deposited to have a periodic structure consisting of alternating SiO_2 and $Ag_xSiO_{2(1-x)}$ layers. The two electron-gun evaporators for evaporating the two solids were turned on and off alternately to make multilayers. The base pressure obtained in IBAD chamber was 6×10^{-6} torr. The growth rate was monitored by an Inficon Quartz Crystal Monitor. The film geometry used for the deposition of $SiO_2/Ag_xSiO_{2(1-x)}$ nano-layers films is shown in Fig.1. The geometry in Fig.1 shows Au contacts on the top and bottom of the multilayers. These two Au contacts were used in the Seebeck coefficient measurements. The electrical conductivity was measured by the Van der Pauw system and the thermal conductivity was measured by the 3ω technique. The electrical conductivity, thermal conductivity and Seebeck coefficient measuremets have been performed at a temperature of $22^{\circ}C$. Detailed information about the 3ω technique may be found in Refs. [6-9]. In order to make nanoclusters in the layers, 5 MeV Si ion bombardments were performed with the Pelletron ion beam accelerator at the Alabama A&M University Center for Irradiation of Materials (AAMU-CIM). The energy of the bombarding Si ions was chosen by the SRIM simulation software (SRIM). The fluences used for the bombardment were $1x10^{13}ions/cm^2$, $5x10^{13}ions/cm^2$, and $1x10^{14}ions/cm^2$. Rutherford Backscattering Spectrometry (RBS) was performed using 2.1 MeV He^+ ions with the particle detector placed at 170 degrees from the incident beam to monitor the film thickness and stoichiometry before and after 5 MeV Si ion bombardment [10, 11].

.

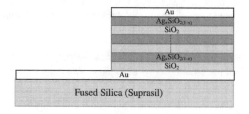

Fig. 1. Geometry of sample from the cross-section.

3.RESULTS AND DISCUSSION

Fig. 2. shows RBS spectrum and RUMP simulation of 50 periodic nano-layers of $SiO_2/Ag_xSiO_{2(1-x)}$ films on a Glassy Polymeric Carbon (GPC) substrate when the sample is at the normal angle. Each element which was used in the deposition is revealed in the RBS spectrum.

RUMP simulation [12] was used to approximate the layer thickness of about 10nm. The total thickness is 486 nm with 50 layers.

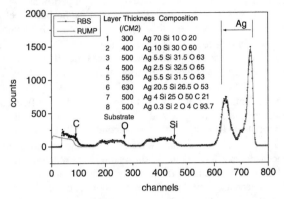

Fig. 2. He RBS spectrum and RUMP simulation for $SiO_2/Ag_xSiO_{2(1-x)}$ nano-layered films on GPC substrate.

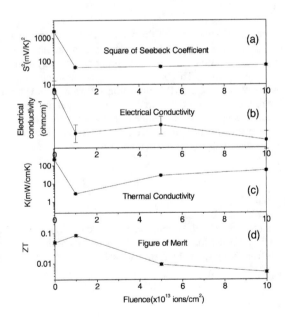

Fig.3. Thermoelectric properties of $SiO_2/Ag_xSiO_{2(1-x)}$ nano-layered films.

Fig.2 belongs to the sample which was unbombarded one. The RBS and RUMP studies for the bombarded films will continue in the future studies to show the effects of the bombardment on the film stoichiometry. Fig. 3 shows the thermoelectric properties of 50 periodic nano-layers of $SiO_2/Ag_xSiO_{2(1-x)}$ films. Figure 3a. shows the square of the Seebeck Coefficient. The graph shows an initial downward slope from the virgin case when the first bombardment fluence of $1x10^{13} ions/cm^2$ was introduced. After the fluence of $1x10^{13} ions/cm^2$, the values are increasing slightly as the fluence increases. The fluence has very little effect on the Seebeck Coefficient. Figure 3b. shows the electrical conductivity as a function of the fluence. Fig. 3b shows almost a periodic behavior of the electrical conductivity. The electrical conductivity starts to decrease from the virgin case until the fluence of $1x10^{13} ions/cm^2$ and as the fluence increases the electrical conductivity values increase and then decreases at almost at the same rate. Figure 3c.shows the thermal conductivity. The thermal conductivity takes an initial drop at the initial fluence of $1x10^{13} ions/cm^2$. As the fluence increases, the thermal conductivity shows an increase while the fluence is increasing. Figure 3d shows the figure of merit. The interesting feature is the first half of the graph. The intial fluence gives an initial increase of the Figure of Merit of 0.09 followed by a decrease until the half way point of the graph. The Figure of Merit continues to decrease with the increasing ion fluence after the fluence of $5x10^{13} ions/cm^2$.

4.CONCLUSION

We have observed the effects of the ion fluence on the thermoelectric properties of the $SiO_2/Ag_xSiO_{2(1-x)}$ alternating layers. The data clearly show that the thermoelectric properties are positively impacted at the initial fluences. The properties quickly degrade or demonstrate limited response to increasing fluence. This behavior suggests ion straggling or damage due to the increasing fluence. We used the SRIM simulation software (SRIM) for determining energy for the system. One of the reason for the degradation and limited response might be the inhomogeneity and change in the stoichiometry in the layers due to ion beam assisted deposition. We could prepare more accurate films to see a better response from our samples. Future studies can provide improvements by focusing on the low ion fluences and exploiting the initial changes lower fluences.

Acknowledgement

Research sponsored by the Center for Irradiation of Materials, Alabama A&M University and by the AAMURI Center for Advanced Propulsion Materials under the contract number NNM06AA12A from NASA, and by National Science Foundation under Grant No. EPS-0447675.

References

1. G. D. Mahan, Solid State Physics, 51, 81 (1998).
2. L.D. Hicks and M.S. Dresselhaus, Physics Review B, 47, 12727 (1993).

3.B C. -K. Huang, J. R. Lim, J. Herman, M. A. Ryan, J. -P. Fleural, N. V. Myung, Electrochemical Acta 50, 4371(2005).

4. T.M. Tritt, ed., Recent Trends in Thermoelectrics, in Semiconductors and Semimetals,71, (2001).

5. Z. Xiao, R.L. Zimmerman, L.R. Holland, B. Zheng, C. I. Muntele, D. Ila, Nuc. Instr. and Meth. B 242,201–204 (2006).

6. L. R. Holland, R. C. Simith, J. Apl. Phys. 37, 4528(1966).

7. D. G. Cahill, M. Katiyar, J. R. Abelson, Phys. Rev.B 50, 6077 (1994).

8. T. B. Tasciuc, A.R. Kumar, G. Chen, Rev. Sci. Instrum. 72, 2139 (2001).

9. L. Lu, W.Yi, D. L. Zhang, Rev. Sci. Instrum. 72 ,2996 (2001).

10. J. F. Ziegler, J. P. Biersack, U. Littmark, The Stopping Range of Ions in solids, Pergamon Press, Newyork, 1985.

11. W. K. Chu, J. W. Mayer, M. -A. Nicolet, Backscattering Spectrometry, Academic Press, New York, 1978.

12. L. R. Doolittle, M. O. Thompson, RUMP, Computer Graphics Service, 2002.

Mater. Res. Soc. Symp. Proc. Vol. 1020 © 2007 Materials Research Society   1020-GG07-22

Effects of MeV Si Ion Bombardments on the Properties of Nano-Layers of SiO2/SiO2+Zn4Sb3

S. Budak[1], S. Guner[1], C. Muntele[1], C. C. Smith[2], R. L. Zimmerman[1], and D. ILA[1]
[1]Center for Irradiation of Materials, Alabama A&M University, 4900 Meridian Street, PO Box 1447, Normal, AL, 35762
[2]MSFC, NASA, MSFC, Huntsville, AL, 35812

Abstract

We prepared 8 periodic nano-layers of $SiO_2/SiO_2+Zn_4Sb_3$. The deposited multi-layer films have a periodic structure consisting of alternating layers where each layer is between 1-10 nm thick. The purpose of this research is to generate nanolayers of nanostructures of Zn_4Sb_3 with SiO_2 as host and as buffer layer using a combination of co-deposition and MeV ion bombardment taking advantage of the electronics energy deposited in the MeV ion track due to ionization in order to nucleate nanostructures. The electrical and thermal properties of the layered structures were studied before and after bombardment by 5 MeV Si ions at various fluences to form nanostructures in layers of SiO_2 containing Zn_4Sb_3. Rutherford Backscattering Spectrometry (RBS) was used to monitor the stoichiometry before and after MeV bombardments.

Keywords: Ion bombardment, Thermoelectric properties, Rutherford backscattering, Van der Pauw method, 3ω method, Seebeck coefficient, Figure of merit

*Corresponding author: D. ILA; Tel.: 256-372-5866; Fax: 256-372-5868;   Email: ila@cim.aamu.edu

1. INTRODUCTION

The single phase β-Zn_4Sb_3 is one of the high performance thermoelectric materials and still attracts great interest due to its convenience primarily on increasing the figure of merit, ZT. The performance of any thermoelectric substance is evaluated by the dimensionless figure of merit expression $ZT = S^2\sigma T / \kappa$, where S, σ, κ and T are the Seebeck coefficient (V/K), electrical conductivity ($\Omega^{-1}m^{-1}$), thermal conductivity (W/m-K) and absolute temperature (K), respectively [1]. In the bulk form β-Zn_4Sb_3, has a high figures of merit (ZT) between 450 and 670 K and a maximum of about 1.3 at a temperature of 670 K while its room temperature ZT is equal to 0.2 [2]. In comparison to bulk form of substances, mono or multilayer thin film fabrication provides a potential to increase the thermoelectric properties of materials. Advantages of reduced size include enhancement of the density of states which increases the Seebeck coefficient [3] and lowers the thermal conductivity [4], both effects increase the ZT. In a study performed by Zhang et al on Zn_4Sb_3 thin films prepared by magnetron sputtering technique, a ZT of 1.2 at 460 K has been obtained for a 349 nm thick Zn_4Sb_3 film specimen [5]. It is also reported that the low electrical resistivity and high Seebeck coefficient can be achieved simultaneously in a film specimen with properly controlled thickness and microstructure. A study on the thermoelectric properties of $Ni_{45}Cu_{55}$ alloy is consistent with this report and proves that the dispersed microstructures of SiO_2 increase the Seebeck coefficient of the material [6].

There are also multilayer thin film investigations performed by Budak *et al* on the MeV Si ion bombardment effects on thermoelectric properties of sequentially deposited $SiO_2/Au_xSiO_{2(1-x)}$ nanolayers [7]. It is reported that the electronic energy deposited due to ionization by a MeV Si beam along ion tracks produces nano-scale structures, which disrupt and confine phonon transmission reducing thermal conductivity, increasing electron density of states such as to increase Seebeck coefficient and the electrical conductivity, thus increasing the figure of merit. In this study, we report the growth of nano-layers of $SiO_2/SiO_2+Zn_4Sb_3$ films on the silica substrates using an ion-beam assisted deposition (IBAD) system, and high energy Si ion bombardment of the films for reducing thermal conductivity and increasing electrical conductivity.

2. EXPERIMENTAL

We have grown nano-layers of SiO_2 and $SiO_2+Zn_4Sb_3$ thin films on silica substrates using Ion Beam Assisted deposition (IBAD) system. The multilayer films were sequentially deposited to have a periodic structure of alternating SiO_2 and $SiO_2+Zn_4Sb_3$ layers. The deposited multi-layer films have alternating thickness of 1-10 nm. One of the electron guns that evaporated the Zn_4Sb_3 was turned on and off alternately to make multilayers. An INFICON deposition monitor controlled the thickness of the layers. The film geometry used in this study is shown in Fig.1. In Fig.1 Au contacts are shown on the top and bottom of the multilayers. These contacts were used in the Seebeck coefficient measurement system.

The electrical conductivity was measured by the Van der Pauw system and the thermal conductivity was measured by the 3ω technique. All the measurements for ZT were performed at room temperature. Detailed information about the 3ω technique may be found in ref. [8]. The 5 MeV Si ion bombardments were performed by the Pelletron ion beam accelerator at the Center for Irradiation of Materials at Alabama A&M University (AAMU-CIM).

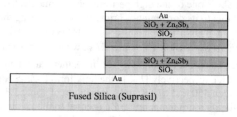

Fig. 1. Geometry of sample from the cross-section.

The energy of the bombarding Si ions was chosen by the SRIM simulation software. The three different fluences used for the bombardment were $1x10^{14} ions/cm^2$, $5x10^{14} ions/cm^2$, $1x10^{15} ions/cm^2$. The RUMP simulation [9]. software was used to analyze the Rutherford backscattering (RBS) spectra for the element concentrations and the thickness of the films.

3. RESULTS AND DISCUSSION

Fig. 2 shows the He RBS spectra of 8 periodic nano-layers of $SiO_2/SiO_2+Zn_4Sb_3$ where the multilayer films were grown on glassy polymeric carbon (GPC) substrates. The quantity of each element is revealed in the RBS spectrum. RUMP simulation gave the thickness of 513 nm and the stoichiometry of 1:1:1:3 for Zn, Sb, SiO2 and C, respectively.

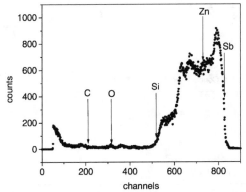

Fig. 2. He RBS spectrum for $SiO_2/SiO_2+Zn_4Sb_3$ nano-layered thin films on GPC substrate.

The room temperature thermoelectric properties of 8 periodic nanolayers of SiO_2 and $SiO_2+Zn_4Sb_3$ superlattice films are shown in the Fig. 3. The changes of the square of the Seebeck coefficient, the electrical conductivity and the thermal conductivity as a function of the fluence of the bombardment are seen in the Fig. 3a, 3b and 3c, respectively. The remarkable increments for each thermoelectric property were observed at the fluence of $1x10^{14} ions/cm^2$. When the fluence values are $5x10^{14} ions/cm^2$ and $1x10^{15} ions/cm^2$, the electrical and thermal conductivities decreased while the Seebeck coefficient was not affected significantly. The increase in the Seebeck coefficient and electrical conductivity is desired while a decrease is preferred for the thermal conductivity. By using the general expression for the dimensionless figure of merit, the changes in this property were calculated and are shown in Fig. 3d. ZT exhibits the maximum increase for the fluence of $1x10^{14} ions/cm^2$ and that enhancement decreases with increasing fluence. Here, we are able to say that when the suitable amount of fluence is chosen, the ion bombardment increases the figure of merit more in the room temperature conditions. Studies will continue on these materials systems to reach the higher ZT.

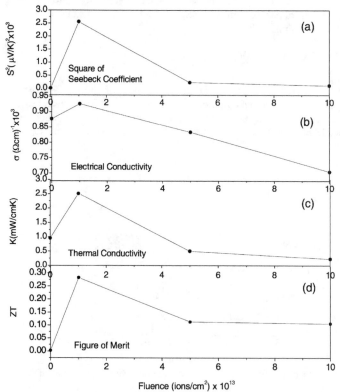

Fig.3. Thermoelectric properties of 8 periodic $SiO_2/SiO_2+Zn_4Sb_3$ nano-layers films.

4. CONCLUSION

We have grown SiO_2 and $SiO_2+Zn_4Sb_3$ thin films on silica substrates using IBAD technique. The multilayer films were sequentially deposited to have a periodic structure consisting of alternating SiO_2 and $SiO_2+Zn_4Sb_3$ layers. The RBS results show that the periodically deposited films have alternating layers 1-10 nm thick. The high energy ion bombardment produce nanostructures and modify the property of thin films, resulting in lower thermal conductivity, higher Seebeck coefficient and electrical conductivity. In our multilayer films, the MeV Si ion bombardment did not cause all these desired results for the same value of fluence. However when the suitable fluence is chosen, there is an increment in the ZT.

Acknowledgement

Research sponsored by the Center for Irradiation of Materials, Alabama A&M University and by the AAMURI Center for Advanced Propulsion Materials under the contract number NNM06AA12A from NASA, and by National Science Foundation under Grant No. EPS-0447675.

References

1. E. Chalfin, H. Lu, R. Dieckmann, Solid State Ionics, 178, 447- 456 (2007).
2. T. Caillat, J.-P. Fleurial and A. Borshchevsky, J. Phys. Chem. Solids, 58, 7, 119-1125. (1997).
3. F. J. DiSalvo, Science 285,703(1999).
4. T. M. Tritt, Science 283, 804 (1999).
5. L. T. Zang, M. Tsutsui, K. Ito, M. Yamaguchi, Thin Solid Films 443, 84-90 (2003).
6. H. Muta, K. Kurosaki, M. Uno, S. Yamanaka, J. Alloys Comp. 359, 326-329 (2003)
7. S. Budak, C. Muntele, B. Zheng, D. Ila, Nucl. Instr. and Meth. in Phys. Res. B (2007), (In press).
8. Chris Dames and Gang Chen, Rev. Sci. Instrum. 76 ,124902 (2005).
9. L. R. Doolittle, M. O. Thompson, RUMP, Computer Graphics Service, 2002.

Mater. Res. Soc. Symp. Proc. Vol. 1020 © 2007 Materials Research Society 1020-GG07-24

High Resolution XRD Studies of Ion Beam Irradiated InGaAs/InP Multi Quantum Wells

Dhamodaran S[1], Sathish N[1], Anand P Pathak[1], Andrzej Turos[2], Devesh K Avasthi[3], and Brij M Arora[4]

[1]School of Physics, University of Hyderabad, Central University (P.O), Hyderabad, 500 046, India
[2]Institute of Electronic Materials Technology, Warsaw, 01-919, Poland
[3]Inter University Accelerator Centre, New Delhi, 110 057, India
[4]Tata Institute of Fundamental Research, Mumbai, 400 005, India

ABSTRACT

To investigate the interface modifications of multi quantum wells InGaAs/InP grown by metal organic chemical vapor deposition have been irradiated using swift heavy ions. Irradiation has been performed using 150 MeV Ag^{12+} and 200 MeV Au^{13+} ions. Both as-grown and irradiated samples were subjected to rapid thermal annealing at 500 and 700^0C for 60s. As-grown, irradiated and annealed samples were subjected to high resolution x-ray diffraction studies. Both symmetric and asymmetric scans were analyzed. The as-grown and Ag ion irradiated samples show sharp and highly ordered satellite peaks whereas, the Au ion irradiated samples show broad and low intense peaks. The higher order satellite peaks of the annealed samples vanished with increase of annealing temperature from 500 to 700^0C, indicating mixing induced interfacial disorder. Annealing of irradiated samples show higher mixing and disorder and no higher order satellite peaks were observed. In comparison with as-grown samples no strain was observed after high temperature annealing of as-grown samples. Strain values calculated from the X-ray studies indicate that the irradiated samples have higher strain 0.2041% which has been reduced as a function of annealing as 0.1704% and 0.1509%. This indicates that the annealing induced mixing occurs maintaining the lattice parameter close to that of the substrate. In similar samples such results have been observed and reported in literature. The effect of electronic energy loss for interface mixing has been discussed in detail. The role of incident ion fluence in combination with the electronic energy loss has also been discussed.

INTRODUCTION

Growth of InGaAs/InP using various epitaxial techniques has been reported. To date, the material that has been used most widely as an absorber is $In_{0.53}Ga_{0.47}As$. The two important characteristics that contribute to the use of this alloy are that (i) it exhibits strong photoresponse in the wavelength range from 1.0 to 1.6μm and (ii) high quality, lattice matched epitaxial layers can be grown on InP. The stability of such heterostructures is important from the device performance point of view. The composition and layer thickness are the parameters used to tune various physical properties. Apart from the lattice matched composition, strain heterostructures of InGaAs/InP are also useful in several applications. In strained heterostructures critical layer thickness is important since beyond this critical thickness defects are generated which deteriorate the device performance. Critical layer thickness has been studied by many groups particularly for highly strained structures [1]. It has been reported that nonequilibrium conditions achieved with Molecular Beam Epitaxy (MBE) generally lead to experimental critical thickness which are very much dependent on the growth conditions [2, 3]. Thermal stability of MBE grown InGaAs/InP samples have also been studied [4]. Metal Organic Chemical Vapor Deposition (MOCVD) grown InGaAs/InP structures have also been studied using various characterization techniques to understand the mismatch strain on the bandgap [5]. Strain relaxation in both compressive and tensile strain has been studied and found that the compressive strain samples were morphologically stable and the layers under tension developed facets [6]. Apart from these studies, bandgap engineering has been investigated by interface mixing using several methods. Impurity free and ion implantation induced mixing of InGaAs/InP QWs have been investigated [7]. Mixing of InGaAs/InP due to high temperature annealing has been studied by Bollet et.al [8]. Influence of cap layer on implantation induced mixing has been carried out by Carmody et.al [9]. From the characterization point of view both as-grown and implanted samples have been studied in detail using the advantages of HRXRD. Both theory and experiments were carried out in detail for InGaAs/InP grown by VPE [10]. Misfit strain using HRXRD and wafer curvature method has been reported. Macrander et.al [11] gave a correlation between the carrier concentration and the X-ray line widths for InGaAs/InP. Ryu et.al studied the diffusion coefficients of cations and anions in InGaAs/InP MQWs due to rapid thermal annealing [12]. Kuphal et.al [13] studied the relation between luminescence wavelength and lattice mismatch. They also established a relation between peak positions of PL spectra at room temperature to the composition x of the $In_xGa_{1-x}As/InP$ layer. Emerson et.al [14] investigated short period superlattices using HRXRD and Raman techniques. They compared experimental and simulated HRXRD scans for the confirmation of obtained Raman results. Compositional gradients in InGaAs/InP structures using HRXRD have also been studied [15, 16]. Interface mixing of as-grown samples have been studied by HRXRD with high accuracy [17, 18]. Using HRXRD the nature of strain was found to be compressive at the lower interface and tensile at the upper interface for the growth conditions used. Positively mismatched InAsP and negatively mismatched InGaAsP inter-layers of higher absolute value are assumed to be responsible for the strain distribution. Cornet et.al studied interfacial properties of InGaAs/InP lattice matched MQWs using HRXRD [19]. Interfacial mixing during growth has been studied in detail and compared with simulation results. High precision of about one monolayer has also been found. In our case the layer composition of indium is x=0.55 which is compressively strained layer. Due to the close lattice parameter values we designate as-grown sample as nearly lattice matched structure. To our knowledge high energy irradiation (energy > 1 MeV/nucleon) has not been utilized for the bandgap modification of semiconductor heterostructures. Recently our group has demonstrated the strain modification in a lattice matched heterostructure using high energy irradiations [20]. Here we present new results concerning high energy irradiation and annealing effects on $In_{0.55}Ga_{0.45}As/InP$ MQWs characterized by HRXRD.

EXPERIMENTAL DETAILS

The InGaAs sample was grown on (001) oriented semi-insulating InP substrates by MOCVD. The MQWs consist of 15 periods of InGaAs/InP with a nominal layer thickness of 20nm each. The Indium composition of the layer have been chosen to be x=0.55 which is nearly lattice matched to the InP substrate. The irradiation was performed at room temperature by 150 MeV Ag^{12+} ions with a fluence of 1×10^{13} ions/cm^2 and 200 MeV Au^{13+} with a fluence of 7×10^{12} ions/cm^2 from NEC 15 MV pelletron accelerator. The ion beam was magnetically scanned over a 1x1 cm^2 area on the sample surface for uniform irradiation. A low beam current was maintained to avoid heating of the samples and the samples were oriented at an angle of about 5^0 with respect to the beam axis to minimize channeling during irradiation. High resolution XRD experiments have been performed using the Philips X'Pert system with a Cu K_α radiation set in point focus mode. System has a channel cut, four crystal Ge(022) (Bartels-type) monochromator for Cu $K\alpha_1$ X-ray beam of divergence 12 arc sec in the scattering plane and a 2^0 open detector. Profiles of symmetric and asymmetric reflections were recorded in $\omega/2\theta$ scans after optimizing the tilt and azimuthal angles. The samples were subjected to Rapid Thermal Annealing (RTA) in order to minimize the irradiation induced damages. As-grown and high energy irradiated InGaAs/InP multi quantum wells were annealed at 500 and 700^0C for 60s in N_2 (1000 SCCM) atmosphere using RXV6 RTP system (AET Thermal Inc). Polished InP wafers were used for proximity capping and the MQWs were placed upside down during annealing. The sample IDs and details are given in Table.1 and the IDs have been used in the text and figures.

Table.1: Sample IDs and details

S.No	Sample ID	Details	Strain (%)
1	M-U	As-grown In$_{0.55}$Ga$_{0.45}$As/InP MQW of 15 periods	0.1315
2	M-I1	150MeV Ag^{12+} ion irradiated with a fluence of 1×10^{13} ions/cm^2	0.2041
3	M-I2	200MeV Au^{13+} ion irradiated with a fluence of 7×10^{12} ions/cm^2	0.1902
4	M-U-A5	M-U sample rapid thermal annealed at 500^{0}C for 60 sec in N$_2$ atmosphere.	~ M-U Sample
5	M-U-A7	M-U sample rapid thermal annealed at700^{0}C for 60 sec in N$_2$ atmosphere.	~ M-U Sample
6	M-I1-A5	M-I1 sample rapid thermal annealed at 500^{0}C for 60 sec in N$_2$ atmosphere.	0.1704
7	M-I1-A7	M-I1 sample rapid thermal annealed at 700^{0}C for 60 sec in N$_2$ atmosphere.	0.1509

RESULTS AND DISCUSSION

The slight compositional variation from the lattice matched structure of InGaAs on InP results in a compressive strain induced in the InGaAs layer. The calculated compressive strain is approximately 0.1315% in the layer. The observation of even-order satellite peaks is an indication of strain and/or non-abrupt interface. The mean superlattice period is calculated from the measured angular separation between the satellite peaks given by,

$$2\Lambda(\sin\theta_n - \sin\theta_0) = \pm n\lambda \qquad (1)$$

where θ_n is the diffraction angle of the order n, θ_0 is the angle of zeroth order peak, λ is the Cu K_α X-ray wavelength of 1.541Å, Λ is the SL period. The calculated value of Λ from (004) scan in Fig.1 is found to be 40.2$\pm$0.1 nm for M-U (as-grown) samples which is close to the nominal value (40nm). Fig.1 shows the (004) scans of M-U and M-I1 (Ag ion irradiated) samples. The inset shows the highly ordered satellite peaks of the M-U sample indicating a good interface quality. Simulation of the HRXRD (Fig.2) has been carried out using the dynamical theory based Philips Epitaxy software. Composition and thickness of the layer has been optimized by a trail and error method starting from the nominal values until a satisfactory fit is observed as explained in Ref. [21]. The simulated scan matches reasonably well with the experimental one for optimized thickness and composition of the layers as $In_{0.55}Ga_{0.45}As$ (20nm)/InP (20nm); this is equal to the nominal ones indicating good quality of the interfaces. Comparably intense and ordered peaks are observed for M-I1 sample with slight broadening of the satellite peaks. Apart from the broadening the shift observed in the satellite peaks of the M-I1 sample with respect to the M-U sample is due to irradiation induced compressive strain in the layer.

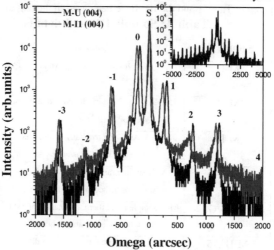

Figure.1: (004) HRXRD scans of M-U and M-I1 samples
(Inset: full spectra of M-U).

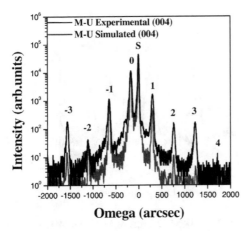

Figure.2: Comparison of simulated and experimental (004) HRXRD scans of M-U

Fig.3 shows the (224)L scans the M-U and M-I1 sample, where a clear shift in the satellite peaks indicates the strain, yet the in-plane strain was found to be negligible. The samples were subjected to RTA to reduce the point defects created due to irradiation. Annealing temperatures of 500 and 700°C for 60s were chosen and both M-U and M-I1 samples were annealed. As a function of annealing temperature the interface mixing induced interfacial disorder is observed from the vanishing satellite peaks. The 500°C annealed sample shows broadening of the satellite peaks but the peak positions were almost the same indicating negligible strain due to annealing. The 700°C annealed samples indicate a reduction in the intensity and broadening of the satellite peaks. A small peak shift is also observed for this sample indicating strain in the sample. Fig.4 shows the comparison of (004) HRXRD scans for M-U, M-U-A7 (As-grown annealed at 700°C) and M-I1-A7 samples. Comparing the annealing effects between these samples, irradiated samples show stronger interface degradation and higher strain induced in the layer. The comparison of strain values of M-I1 and M-I1-A7 sample with M-U sample shows a reduction due to annealing. The average compressive strain of 0.2041% in M-I1 sample has reduced to 0.1704% and further to 0.1509% as a function of annealing. This would mean that the thermal energy provided results in preferential diffusion of elements to maintain the lattice parameter matched to the substrate. Preferential diffusion and mixing by maintaining the lattice parameter close to that of the substrate have been reported [22, 23]. Such results are comparable with the present studies. Fig.5 shows the (004) scans of M-U, M-I1 and M-I2 (Au ion irradiated) samples (shown in different colours). The broad satellite peaks for M-I2 samples indicates huge degradation of the interface. The strain induced in M-I2 sample (0.1902%) is lower than M-I1 but higher than M-I1-A7. In spite of the lower electronic energy loss of 150MeV Ag than 200MeV Au, higher strain is observed in M-I1. It is to be noted that the sample M-I1 has been irradiated with higher fluence (1×10^{13} ions/cm^2) than compared with the M-I2 (7×10^{12} ions/cm^2) the strain variation indicates that the fluence might play a bigger role than the energy loss.

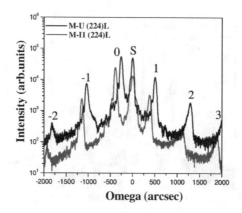

Figure.3: (224) HRXRD scans of M-U and M-I1 samples.

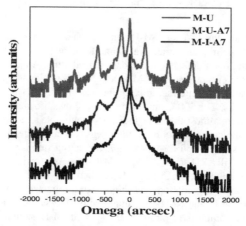

Figure.4: (004) HRXRD scans of M-U, M-U-A7 and M-I1-A7 MQWs.

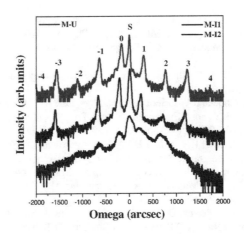

Figure.5: (004) scans of M-U, M-I1 and M-I2 MQWs.

CONCLUSION

Interface modification of MQWs by heavy ion irradiation has been investigated using HRXRD. Results of the as-grown and irradiated samples after annealing have been compared and found that the average compressive strain induced by irradiation reduces upon subsequent annealing. Preferential diffusion and mixing to maintain the lattice parameter of the layer close to that of the substrate have been attributed to these observations. Fluence gives higher mixing efficiencies than the electronic energy loss of the ions. This is observed from the strain values calculated for the Ag and Au ion irradiated samples. Such interfacial modifications are useful in tailoring the material properties spatially.

ACKNOWLEDGEMENTS

SD is grateful to CSIR, India for the award of Senior Research Fellowship. NS thanks SSPL, Delhi for fellowship through a project granted to APP. We acknowledge the support by Indo Polish joint scientific and technological cooperation programme. We are grateful to Mr. G Sai Saravanan for help in annealing the samples.

REFERENCES

1. M Gendry, V Drouot, C Santinelli and G Hollinger Appl. Phys. Lett. 60, 2249 (1992).
2. P R Berger, K Chang, P Bhattacharya, J Singh, and K K Bajaj, Appl. Phys. Lett. 53, 684 (1988).
3. M J Ekenstedt, S M Wang and T G Andersson, Appl. Phys. Lett. 58, 854 (1991).
4. H Temkin, S N G Chu, M B Panish and R A Logan, Appl. Phys. Lett. 50, 956 (1987).
5. I C Bassignana, C J Miner and N Puetz, J. Appl. Phys. 65, 4299 (1989).
6. T Okada, G C Weatherly and D W McComb, J. Appl. Phys. 81, 2185 (1997).
7. L V Dao, M Gal, C Carmody, H H Tan and C Jagadish, J. Appl. Phys. 88, 5252 (2000).

8. F Bollet, W P Gillin, M Hopkinson and R Gwilliam, J. Appl. Phys. 93, 3881 (2003).
9. C Carmody, H H Tan and C Jagadish, J. Appl. Phys. 93, 4468 (2003).
10. S N G Chu, A T Macrander, K E Strege and W D Johnston Jr, J. Appl. Phys. 57, 249 (1984).
11. A T Macrander, S N G Chu, K E Strege, A F Bloemeke and W D Johnston Jr, Appl. Phys. Lett. 44, 615 (1984).
12. S W Ryu, B D Choe and W G Jeong, Appl. Phys. Lett. 71, 1670 (1997).
13. E Kuphal, A Pocker and A Eisenbach, J. Appl. Phys. 73, 4599 (1993).
14. D T Emerson and J R Shealy, Appl. Phys. Lett. 69, 383 (1996).
15. M A G Halliwell, M H Lyons, M J Hill, J. Cryst. Growth. 68, 523 (1984).
16. S Bensoussan, C Malgrange and M S Simkin, J. Appl. Cryst. 20, 222 (1987).
17. A Krost, J Bohrer, H Roehle and G Bauer, Appl. Phys. Lett. 64, 469 (1994).
18. J M Vandenberg, A T Macrander, R A Hamm and M B Panish, Phys. Rev. B 44, 3991 (1991).
19. D M Cornet R R LaPierrea, D Comedi and Y A Pusep, J. Appl. Phys. 100, 43518 (2006).
20. S V S Nageswara Rao, A P Pathak, A M Siddiqui, D K Avasthi, C Muntele, D Ila, B N Dev, R Muralidharan, F Eichhorn, R Groetzschel, A Turos, Nucl. Instr. and Meth. B 212 (2003) 442.
21. S Dhamodaran, N Sathish, A P Pathak, S A Khan, D K Avasthi, T Srinivasan, R Muralidharan and B M Arora, Nucl. Instr. and Meth. B 256 (2007) 260.
22. S J Yu, H Asahi, S Emura, S Gonda and K Nakashima, J. Appl. Phys. 70, 204 (1991).
23. S J Yu, A J Takizawa, K Asami, S Emura, S Gonda, H Kubo, C Hamaguchi and Y Hirayama, J. Vac. Sci and Technol. B 9, 2683 (1991).

Mater. Res. Soc. Symp. Proc. Vol. 1020 © 2007 Materials Research Society   1020-GG08-01

Focused Ion Beam Fabrication of Individual Carbon Nanotube Devices

Lee Chow[1], and Guangyu Chai[2]
[1]Physics, University of Central Florida, 4000 Central Florida Blvd., Orlando, FL, 32816-2385
[2]Apollo Technologies Inc., Orlando, FL, 32750

ABSTRACT

Focused ion beam (FIB) techniques have found many applications in nanoscience and nanotechnology applications in recent years. However, not much work has been done using FIB to fabricate carbon nanotube devices. This is mainly due to the fact that carbon nanotubes are very fragile and energetic ion beam from FIB can easily damage the carbon nanotubes. Here we report the fabrication of carbon nanotube (CNT) devices, including electron field emitters, atomic force microscope tips, and nano-pores for biomedical applications. This is made possible by a unique, coaxial configuration consisting of a CNT embedded in a graphitic carbon coating, which was developed by us for FIB processing of carbon nanotubes. The CNT-based atomic force microscope tip has been demonstrated. The electron field emission from the tip and the side wall of CNT will be discussed. We will also report the fabrication of a multiwall carbon nanotube nanopore for future applications.

INTRODUCTION

Since its discovery [1] in 1991, carbon nanotube (CNT) attracted intense interest because of their potential applications in nanotechnology [2-6]. All these applications are due to the unique properties of CNTs, among which are, high aspect ratio, large Young's modulus [7], chemical inertness, and excellent field emission properties [8]. There are many demonstrated and proposed applications of carbon nanotube devices. In general, we can classify them into two categories: applications that required large quantities of CNTs, and applications that only required one single CNT. Here in this paper, we will concentrate on the applications that only requires one single CNT in each device. In particular, the following three carbon nanotube devices will be considered: carbon nanotubes as electron sources for electron microscopes, carbon nanotubes as sharp probe tips for the atomic force microscope, and carbon nanotubes as nano-pore for biological applications.

Focused ion beam (FIB) [9, 11] is a very versatile technique to remove or to deposit materials. Because of its ability to focus an ion beam to a spot size of a few nanometers, FIB has found numerous applications [12] in nanofabrication. The focused ion beam technology is based on liquid metal ion source and ion optics technology developed in the late '70 and early '80. The two most important functions of an FIB system are its ability to fabricate nanometer size structures and to deposit minute amounts of materials with nanometer precision.

Recently, researchers started to apply FIB techniques to CNT-related applications. The initial efforts involved applying FIB to modify and manipulate CNTs directly with Ga+ ion beams [13-16]. However, due to the high impact momentum of Ga$^+$, the carbon atoms are easily knocked out from their lattice locations and structural defects were induced inside the CNTs. These defects are unavoidable when CNTs are directly exposed to a Ga$^+$ ion beam, and weaken the electrical and mechanical properties of the CNTs which are essential for CNT applications. Since the CNT defects generated by FIB appeared to be inevitable, researchers started to use the defects on CNTs intentionally. Some promising applications included using FIB to open capped single-wall carbon nanotubes (SWNT) [17] and applying FIB to generate damaged areas on SWNTs to fabricate quantum dots[18].

EXPERIMENTAL PROCEDURE

We have developed a chemical vapor deposition technique [19] to grow a fiber-protected carbon nanotubes (F-CNT). Our main goal was to grow a configuration of carbon nanotube that would be easy to handle and could withstand the intense ion beam in a focused ion beam instrument. We adopted the following strategy to accomplish the task. We first prepared 30 nm size catalysts on a silicon substrate using evaporation under Helium atomsphere. We then heated our substrate under a 50/50 mixture of Ar and H$_2$ gases to 725 °C to purge the growth chamber and to reduce the Fe/Ni nanoparticles to metallic form. We then turned off H$_2$ gas, and add 10% methane with Ar and continued heating the substrate to 875 °C. During this stage, carbon nanotubes were grown on the silicon substrate by the catalytic vapor growth process. The length of the nanotube depended on the growth duration while the diameter of the nanotube is mainly controlled by the initial catalytic nanoparticle size. We then reduced the methane concentration to 5% of the total mixture and added 5% of H$_2$ and increased the temperature to a level between 900 and 1000 °C for 10 minutes. This step was used to deposit a thin layer of amorphous carbon on the outside wall of the CNTs. Afterward, we increased the temperature to 1000 °C for 30 minutes. During this final step, a graphitic carbon outer layer is formed on top of the thin amorphous carbon. In Figure 1, an SEM micrograph of the F-CNT is shown together with a schematic diagram of the F-CNT.

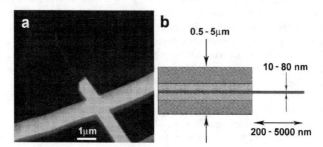

Figure 1. (a) SEM image of the as-grown fiber-protected carbon nanotube. (b) A schematic diagram of the fiber-protected carbon nanotube.

Once the F-CNTs were synthesized, the outer carbon layer was broken mechanically. The carbon nanotube is stronger than the carbon fiber, so it is possible to crush the outer carbon layer and reveal the inner CNT core as shown in Figure 1(a) above. This way, we end up with a co-axial nano-structure with the CNT as the core of micron size carbon fiber which can be easily manipulated with a micro-manipulator under an optical microscope. These F-CNTs were then transferred to FIB chamber using a clean tungsten wire with silver paint at the tip to glue these micron size fibers. It should be noted here that due to the minute size of carbon nanotubes, the nanotubes can be easily damaged by energetic ion beams. We observed a "softening" of the CNTs due to exposure to Ga^+ ion beams in our recent work [20]. It is very desirable to use a dual-beam FIB instrument for the processing of carbon nanotube devices for the above reason. However, our work reported below was done on a single beam FEI Vectra 200 FIB instrument.

FIB fabrication of CNT atomic force microscope (AFM) tip

A tungsten needle micro-manipulator installed inside the FIB chamber is used to assist the fabrication of the CNT AFM tip. We first use the micromanipulator to pick up an intermediate carbon fiber. This step is necessary to give us an extra degree of freedom to manipulate the CNT samples. For each "pick-up" procedure, we first use FIB milling to create a slot on the tungsten needle, we then move the manipulator and intermediate fiber together and deposit Pt metal at the joint to weld the fiber and tungsten needle together (see Figure 2(a)). Afterward, FIB is used to cut the intermediate fiber away from the substrate. The reason to mill a slot before each welding step is to increase the strength of the weld joint. Once we have the intermediate fiber welded on the manipulator, we then use the exact same procedure to pick up the desired fiber that contains the CNT at the end (see figure 2(b)). Next, we cut a deep slot at the tip of a conventional AFM cantilever pyramid with FIB, move the fiber into the slot, weld them together, and cut off the intermediate fiber from the AFM tip (See Figure 2(c)).

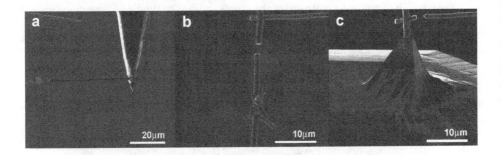

Figure 2.   Secondary electron images of the FIB fabrication process. (a) Using a tungsten needle to pick up a carbon fiber as the intermediate fiber. (b) Using the intermediate fiber to pick up the desired fiber with nanotube at the end of the fiber. (c) Mounting the fiber on the pyramid tip of an AFM cantilever.

In figure 3 below, we show an SEM image of an AFM cantilever with a carbon nanotube positioned at the tip of the pyramid. As can be seen in the image, the micron size fiber with a nanotube tip is bonded to the top of the pyramid tip. The CNT tip is perfectly aligned vertically with the AFM cantilever. A close-up high magnification SEM image on the right shows that this carbon fiber has a diameter about 1.3 μm, the CNT radius is about 20 nm, and the CNT length is about 150 nm.

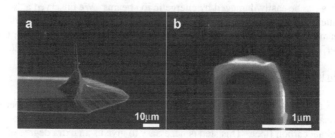

Figure 3. (a) The AFM cantilever with F-CNT tip mounted on the top of the pyramid. (b) A close-up image shows the carbon fiber with its nanotube core.

FIB fabrication of CNT electron field emitter

The fabrication of a CNT electron field emitter [21] is very similar to the fabrication of a CNT AFM tip. We start out with a 200 μm diameter tungsten wire and etch the tip down to a radius of less then 0.1 μm. Then we mounted the etched tungsten tip in the FIB instrument and used the FIB to mill a slot at the end of the tungsten wire. The same procedure described above is used to pick up the carbon fiber which contains the carbon nanotube and mount it on the tip of the tungsten wire as shown in Figure 4(a) below. In Figure 4(b) a close-up SEM image of the carbon nanotube is shown.

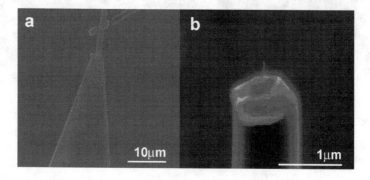

Figure 4. (a) Mounting a fiber-protected CNT on an etched W filament. (b) A close-up view of the tip of carbon fiber showing the CNT at the core of the carbon fiber.

Since one of the major applications for a CNT field emitter will be to act as the electron source for a field emission electron microscope, we also demonstrated that it is possible to fabricate our carbon nanotube field emitter on a conventional tungsten filament SEM source. In figure 5, our carbon nanotube field emission tip is shown on a commercial tungsten filament. Here we used the FIB to fabricate a slot at the apex of the tungsten filament first, and then carbon fiber bearing the CNT tip was mounted in the slot and Pt metal was used to secure the joint, as in the procedure described earlier.

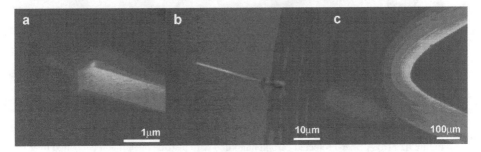

Figure 5.   Secondary electron images of carbon nanotube fabricated using FIB on a conventional tungsten SEM electron source. (a) A close-up view of the carbon nanotube tip on the carbon fiber. (b) A broader view of the carbon fiber at the apex of the W filament. (c) A zoom-out view of W filament with the F-CNT barely seen.

FIB fabrication of CNT nanoaperture/nanopore

Another area of research that we started recently involves using focused ion beam to fabricate carbon nanotube nanoaperure/nanopore. In 2001, Meller et al [22] demonstrated that DNA can be driven through a nanopore with an applied voltage. The authors used a biological nanopore: α-hemolysin. Although biological nanopores are useful, there are a lot of advantages to use solid state nanopores [23-25], such as stability and controllability. In 2002, Siwy et al [26] used a track etching technique to produce a synthetic nanohole on a membrane. The intense interest in nanopores are due to the possibility of using nanopores as a sequencing tool for DNA molecules.

To fabricate a CNT nanopore, we started out with our fiber-protected carbon nanotube lying on a silicon substrate. First, a stripe of Pt metal about 3μm x 20μm was deposited perpendicular to the length of carbon nanotube to protect and secure the carbon fiber. We then used the standard TEM cross-sectional sample preparation technique [27] to lift a cross-section of carbon nanotube and silicon sample. This sample is the size of a typical TEM sample, so it is not difficult to handle. The sample is directly transferred to a TEM grid inside the FIB chamber for future measurements.

In Figure 6(a), a fiber-protected carbon nanotube lies flat on the silicon substrate is shown. In Figure 6(b), the lift-off procedure is half completed in this image. We can see that the Pt metal has been deposited and two large cavities have been milled on both sides of the sample.

In Figure 6(c), the sample has been transferred from the silicon substrate and mounted on a pre-drilled slot on the side of a copper TEM grid. Note that the magnification in this image has been reduced by a factor of 10, so the sample appears smaller.

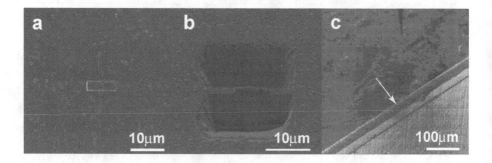

Figure 6. (a) A fiber-protected carbon nanotube with a stripe of Pt metal deposited over it. (b) Standard TEM sample preparation technique is used to lift off a cross-section sample from the substrate. (c) Use the *in-situ* micromanipulator to transfer the sample to a Cu TEM grid for further testing under TEM. The arrow indicates the position of the sample.

The possible exponential improvement in the quantum computation speed has spark intense interest in quantum computer [28]. One possible realization of a solid state quantum computer [29] would be an array of ^{31}P atoms embedded in an isotopically enriched ^{28}Si substrate. To implement the above idea, single ion implantation with nanometer accuracy is required. We want to note here that the CNT nanopore sample prepared above can be used as a nano-aperture for single ion implantation [30-31], or as a channeling device for charged particles [32].

RESULTS and DISCUSSION

CNT atomic microscope tip

The CNT-tipped AFM cantilever fabricated is then mounted into a Nomad scanning probe microscope from Quesant Instruments for testing. A standard optical interference grating pattern is used as a sample for the comparison between our CNT tip and a commercial silicon tip. The results are shown in Figure 7. In Figure 7(a), the image was obtained using our CNT tip, while in Figure 7(b), a new silicon tip was used. Here we can see that the image from CNT tip indeed showed much more details and the size of the grains were much smaller. The image scanned using a new silicon tip in Figure 7(b) showed a much coarser grain size.

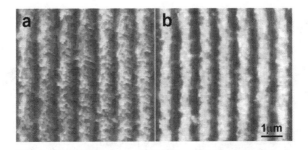

Figure 7. (a) Image scanned by a CNT AFM tip fabricated with focused ion beam. (b) Image
scanned by a conventional silicon AFM tip.

In addition to better spatial resolution, a CNT AFM tip has the advantages of longer
operating lifetime and the ability to image the bottom of a narrow trench [2]. A typical
conventional silicon AFM tip can only last for about 30 minutes before the tip starts to wear out,
while a CNT-based scanning probe tip can last much longer as shown in Figure 8 below.

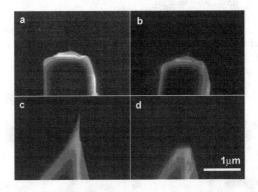

Figure 8. (a) CNT before scan. (b) CNT tip after 2 hr of scan. (c) Si tip before scan. (d) Si tip
after 0.5 hour of scan.

Current voltage characteristic of a CNT electron field emitter

Individual CNT-based electron field emitters fabricated by FIB techniques were tested in
an ultrahigh vacuum chamber [20, 21]. We found that the I-V characteristic curve of the field
emission followed the standard Fowler-Nordheim relationship with a turn-on voltage of 120
volts. A typical I-V curve and a Fowler-Nordheim plot of the electron emission from the CNT
tip are shown in Figure 9 below.

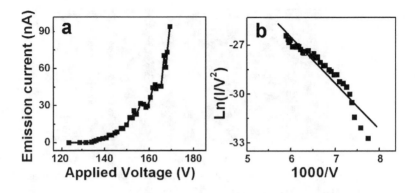

Figure 9. (a) A typical I-V characteristic curve of CNT field emitter. (b) The semi-logarithmic plot shows the Fowler-Nordheim relationship.

TEM testing of the nanopore sample

To check the nanometer size channel of the carbon nanotube sample in Figure 6(c), we used a FEI Tecnai F30 scanning transmission electron microscope. We employed a 120 KeV unfocused electron beam and a CCD detector behind the sample to detect electrons passed through the hollow core of CNT. In Figure 10 below, we can see clearly the image produced by electrons channel through the core of carbon nanotube.

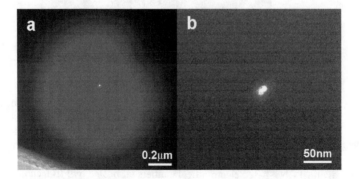

Figure 10. (a) Image of the electrons channel through the CNT hollow core detected by a CCD detector. (b) A magnified image of electrons channel through CNT hollow core. The diameter of the hole is about 12-15 nm.

CONCLUSIONS

In conclusion, in this paper we demonstrated that focused ion beam techniques can be used to fabricate carbon nanotube nano-devices. The FIB fabrication process is aided by the synthesis of a unique fiber-protected carbon nanotube configuration. Three different types of CNT-based nano-devices were demonstrated. We showed that CNT AFM tips can be fabricated with FIB. The device has the advantages of a large aspect ratio and a significantly longer operating lifetime. We also showed that F-CNT electron filed emitter can be fabricated with FIB method. These CNT-based electron field emitters can be used as electron sources in a field emission electron microscope. We also demonstrated the fabrication of nanopores based on carbon nanotubes. The nanopore we constructed has an inner diameter of 12 nm and the length of the nanopore can be adjusted.

ACKNOWLEDGMENTS

The authors would like to acknowledge the contributions from our past and current collaborators: Dr. Dan Zhou, Dr. S. Kleckley, Dr. R. Vanfleet, Dr. Weili Luo, Dr. H. Heinrich, and Dr. T. Schenkel. We also acknowledge Dr. Ted Tessner of FEI company for providing the tungsten filament used in figure 4. Technical support from the staff of UCF's Material Characterization facility and partial financial support from Apollo Tech, Inc. is greatly appreciated.

REFERENCES

1. S. Iijima, Nature **354**, 56 (1991).
2. H. Dai, J. H. Hafner, A. G. Rinzler, D. T. Colbert, and R. E. Smalley, Nature **384**, 147 (1996).
3. W. A. De Heer, A. Chatelain, and D. Ugarte, Science, **270** 1179 (1995).
4. S. J. Trans, A. R. M. Verschuren, and C. Dekker, Nature **393**, 49 (1998).
5. P. G. Collins and A. Zettl, Appl. Phys. Lett. **69,** 1969 (1996).
6. A. Bachtold, P. Hadley, T. Nakanishi, and C. Dekker, Science, **294**, 1317 (2001).
7. M. M. J. Treacy, T. W. Ebbesen, and J. M. Gibson, Nature, **381**, 678 (1996).
8. G. Rinzler, J. H. Hafner, P. Nikolaev, L. Lou, S. G. Kim, D. Tomanek, P. Nordlander, D. T. Colbert, and R. E. Smalley, Science, **269**, 1550 (1995).
9. J. Orloff, M. Utlaut, and L. Swanson, "High Resolution Focused Ion Beams", Kluwer Academic/Plenum Publishers, New York, (2003).
10. L. A. Giannuzzi and F. Stevie, "Introduction to Focused Ion Beams: Theory, Instrumentation, Applications, and Practice", Springer, NY (2005).
11. Wagner, J. P. Levin, J. L. Mauer, P. G. Blauner, S. J. Kirch, and P. Longo, Journal of Vacuum Science & Technology B, **8**, 1557 (1990).
12. A. A. Tseng, Small, **1** 924 (2005).
13. M. S. Raghuveer, P. G. Ganesan, J. D'Arcy-Gall, and G. Ramanath, Appl. Phys. Lett., **84**,4484 (2004).

14. Y.J. Jung, Y. Homma, R. Vajtai, Y. Kobayashi, T. Ogino, and P. M. Ajayan, Nano Letters, **4**, 1109 (2004).

15. Z. Deng, E. Yenilmez, A. Reilein, J. Leu, H. Dai, and K. A. Moler, Appl. Phys. Lett., **88**, 023119 (2006).

16. C. S. Han, J. K. Park, Y. H. Yoon, and Y. H. Shin, Carbon, **44**, 3348 (2006).

17. M. J. Kim, E. Haroz, Y. Wang, H. Shan, N. Nicholas, C. Kittrell, V. V. Moore, Y. Jung, D. Luzzi, R. Wheeler, T. BensonTolle, H. Fan, S. Da, W. Hwang, T. J. Wainerdi, H. Schmidt, R. H. Hauge, and R. E. Smalley, Nano Letters, **7**, 15 (2007).

18. K. Maehashi, H. Ozaki, Y. Ohno, K. Inoue, K. Matsumoto, S. Seki, and S. Tagawa, Appl. Phys. Lett., **90**, 023103 (2007).

19. S. Kleckley, G. Y. Chai, D. Zhou, R. Vanfleet, and L. Chow, Carbon, **41** 833 (2003).

20. G. Chai and L. Chow, Carbon, **45**, 281 (2007).

21. G. Chai, L. Chow, D. Zhou, and S. P. Byahut, Carbon **43**, 2083 (2005).

22. A. Meller, L. Nivon, and D. Branton, Phys. Rev. Lett. **86**, 3435 (2001).

23. C. Schmidt, , M.Mayer, and H. Vogel, Angew. Chem. Int. Edn, **39**, 3137 (2000).

24. N. Fertig, *et al.* , Phys. Rev. E **64**, 040901 (2001).

25. C. Dekker, Nature Nanotechnology, 1, to appear. (2007).

26. Z. Siwy and A. Fulinski, Phys. Rev. Lett. 89, 198103 (2002).

27. Yamaguchi, M. Shibata, and T. Hashinaga, Journal of Vacuum Science & Technology B, **11**, 2016 (1993).

28. C. H. Bennett and D. P. DiVincenzo, Nature **404**, 247 (2000).

29. B. E. Kane, Nature 393, 133 (1998).

30. T. Schenkel, A. Persaud, S. J. Park, J. Nilsson, J. Bokor, J. A. Liddle, R. Keller, R. H. Schneider, D. W. Cheng, and D. E. Humphries, J. Appl. Phys. **94**, 7017 (2003).

31. T. Schenkel, V. Radmilovic, E. A. Stach, S. J. Park, and A. Persaud, J. Vac. Sci. & Tech. B **21**, 2720 (2003).

32. A. V. Krasheninnikov and K. Nordlund, Phys. Rev. **B** 71, 245408 (2005).

Mater. Res. Soc. Symp. Proc. Vol. 1020 © 2007 Materials Research Society   1020-GG03-06

MeV Ion Beam Fabrication of Nanopores

Renato Amaral Minamisawa, Robert Lee Zimmerman, Claudiu Muntele, and Daryush ILA
Physics, Center for Irradiation of Materials, Alabama A&M University, PO Box 1447, Normal, AL, 35762

ABSTRACT

We have used MeV ion beams to fabricate nanopores in Poly(tetrafluorethylene-co-perfluoro-(propyl vinyl ether)) (PFA) fluoropolymer membranes. We have developed an in house system to produce nanopores. Using MeV ion beams we developed a method to produce pores from nanometers to one-micron diameter. A thin film of the PFA polymer was mounted to cover a window to a gas filled chamber and then exposed to a uniformly scanned MeV ion beam masked to define the exposed area. The gas leak rate through the fabricated pores was monitored by an *in situ* RGA system both during and after each bombardment to correlate the leakage with the total area of the pores produced. In this project we used MeV light and heavy ions to best define the pore diameter through each hole and the pore entrance and exit dimensions in the membranes.

INTRODUCTION

Porous membranes using synthetic polymers have been applied in several research methods and devices such as hydrophobic filters for removal of microorganisms and particles from air and other gases, chromatography gases detection [1], biologic detectors etc. PFA, which is a fully fluorinated something with oxygen cross links between chains expressed by the molecular formula $[(CF_2CF_2)_nCF_2C(OR)F]_n$, proved suitable for a large range of these porous membrane applications because it is chemically inert, with high chemical resistance to solvents and has a melting point around 304 °C [2]. The controlled fabrication of pores in PFA in different length scales may provide a material with high quality performance for membranes applications.

Attempting to produce micropores in PFA, Caplan *et al.* [3] have developed in 1997 a method using thermally induced phase separation. However the process is time consuming and makes it difficult to control the size of the fabricated pores. In 2003 Apel *et al.* [4] reported the fabrication of nanometer (~500 nm) holes in poly(ethylene terephthalate) (PET) polymer films using MeV to GeV energy ion track etch technique. However, although the spatial resolution is efficient for biologic membrane transport studies, it is limited to only several applications. To overcome these problems, Li *et al.* and Stein *et al.* [5, 6] have recently proposed a successful ion-sculpted method using a keV focused ion beam (FIB) apparatus to produce pores in the nanometer length scale in thin, isolating silicon nitride solid-state membranes. Although the Si_3N_4 is easier to be manipulated, the high chemical inertness of synthetic polymers is still a desired property.

We report in this article recent studies of pores fabrication at micro and nano scale in PFA thin films using a homemade feedback controlled ion beam system adapted to a MeV Pelletron accelerator. Structural chemical changes in the PFA thin films bombarded with MeV ions will be reported as well as preliminary results of the technique developed to control the pores formation. The nanoporous PFA membranes may be promising for molecular electronics, rapid DNA sequentially strategies and membrane filter technologies.

EXPERIMENTAL

Poly(tetrafluorethylene-co-perfluoro-(propyl vinyl ether)) film (DuPont) with 12.5 μm thickness were prepared in 2×2 cm^2 dimensions square shape.

The homemade feedback controlled ion beam system consists in a He gas reservoir chamber with regulated pressure, having the thin film as a covering window between the accelerator bombardment chamber, where was placed a Residual Gas Analyzer (RGA), and the He gas volume. Figure 1 shows a schematic model of the feedback controlled ion beam system. Specifications of the experimental apparatus will be fully described in further works.

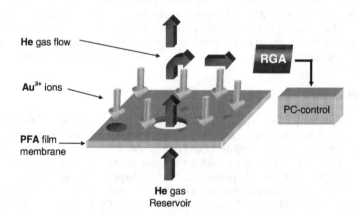

Figure 1. Schematic of the feedback controlled ion beam system.

The gas was restricted inside of a 10 mm^3 reservoir volume (at 760 torr) in order to protect the accelerator vacuum system (at 10^{-7} torr) in case of mechanical fracture of the PFA thin film. The samples were bombarded with a 5 MeV Au^{3+} ion beam, aligned with the He chamber. The ion beam was scanned during bombardment to obtain a homogeneous distribution of ions in the sample. A 400 μm diameter aluminum circular hole and a 10 μm thickness mesh grid with 2000 squares per millimeter in contact with the film were used as holders and ion collimators for the PFA window. At 5 MeV energy, the Au^{3+} ion beam penetrates in the film inducing structural chemical changes that results in bonds breaks and molecules sputtering from the PFA surface. These chemical structural changes in the PFA bombarded films at different fluences were evaluated using a Lab-Ram Micro Raman spectrometer equipped with a He-Ne laser with 632.87 nm excitation wavelength. Micro Raman measurement is surface sensitive in order to acquire scattering only from the effected region of the bombarded samples. The time acquisition was kept constant for the ensemble of sampler.

As soon as the He gas leaks through the pores it is detected by the RGA, the feedback system blocks the beam with a Faraday cup. The pores characterization was made in a Solver P47H Scanning Probe Microscope (NT-MDT) in the tapping mode in order to not damage the sample surface.

RESULTS AND DISCUSSION

Raman analysis is suitable to evaluate the molecular composition of PFA ion damaged areas after bombardment, which may change the physical chemistry properties that are important for membrane applications. Results are shown in Figure 2 where one can notice no changes in the C-F and C-O bonds with peaks around 731.0 and 1381.1 cm^{-1}, respectively, for the samples bombarded at 1×10^{13} fluence.

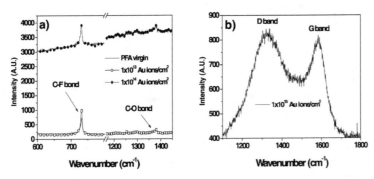

Figure 2. Raman spectra of PFA (a) virgin sample and bombarded at 1×10^{13} and 1×10^{14} ions/cm^2, showing no bond changes after irradiation and (b) 1×10^{15} ions/cm^2 where the C-F and C-O bonds disappear and the D and G bands from graphite-like materials form.

The sample bombarded at 1×10^{14} ions/cm^2 fluence (Figure 2(a)) did not show bond changes although its acquisition yield increased by ten times. We propose that this effect could be attributed due to fluorescence or to surface enhanced Raman scattering (SERS) due to the influence of the implanted Au particles in the PFA surface that formed nanometer sized metal clusters or surface grains [7, 8]. Incident laser photons are absorbed into the metal particles through oscillations of surface electronic charge density named plasmons, which radiation can couple with the polymer molecules in close proximity and provide an efficient pathway to transfer energy to their vibrational modes, and generate the Raman scattering enhancement.

The PFA characteristic C-F and C-O bonds signals disappeared in the sample bombarded at 1×10^{15} ions/cm^2 fluence (Figure 2(b)) giving place to the D and G vibrational modes from amorphous carbon with peaks around 1329 and 1585 cm^{-1}, respectively. The D band is usually assigned to zone centers phonons of the E_{2g} symmetry and the G band to K-point phonons of the A_{1g} symmetry [9]. The 5 MeV Au at 1×10^{14} ions/cm^2 fluence, used to produce the first pores in the 12.5 μm PFA film, did not generate serious damage in the molecular structure of the membrane.

In this accumulated fluence, Au particles sputtered the PFA surface, collimated by the mesh grid, and drilled a set of 5 μm holes through the film. However, the pores fabricated in the bottom surface of PFA were monitored and controlled by the increase in the RGA detector gas flow rate indication. Although a spontaneous diffusion of He through the film is observed before the bombardment, the gas flow rate detected showed clearly changes during the pores formation because the partial pressure of He increases in the chamber. Figure 3(a) shows an Atomic Force Microscopy image of the first pores fabricated in the back of the PFA surface after feedback controlled bombardment of the front face. Along the entire scanned area of the thin film, dispersed pores were observed with diameters randomly ranged between 100 nm to 500 nm. Figure 3(b) shows the AFM scan of a nanopore and it depth profile.

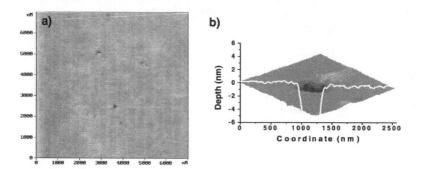

Figure 3. (a) Scanning probe microscope image of an area of the non-bombarded back face of PFA, showing few and dispersed holes, and (b) an image of a 300 nm fabricated pore with the depth profile curve.

Nanopores with 100 nm diameters were also found with the larger ones in the PFA bombarded samples as shown on Figure 4. This nanopore scale size usually requires focused ion beam systems, but our technique made possible to produce it with a scanned ion beam in a fast process. Is important to highlight that the tip dimensions, used for the measurements, limited the spatial resolution of the scanning probe microscopy technique. Smaller nanopores may require TEM technique to be identified.

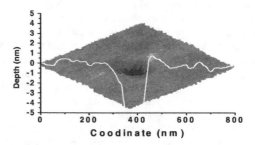

Figure 4. AFM image of a 100 nm diameter nanopores.

CONCLUSIONS

Poly(tetrafluorethylene-co-perfluoro-(propyl vinyl ether)) porous membranes are used for several applications in research and industry. Control in the number and a size of the pores in the synthetic thin film is important for each specific application and may open new fields for nanotechnology studies. The homemade feedback controlled ion beam system developed in the Center for Irradiation of Materials at Alabama A&M University is promising for the nanopores ion beam fabrication and detection. The system consists in the detection of pores formation using the He gas leakage trapped in a reservoir through a bombarded PFA film window. Bombardment using Au ions at 5 MeV energy demonstrate little damage in the chemical structure of the thin film polymer and high sputtering yield for the effective fluence necessary to fabricate the nanopores. Diameter of the nanopores ranges between 100 and 500 nm and are randomly distributed in the back face of the bombarded regions. The measurements were limited by the AFM spatial resolution. This article presented the MeV ion beam fabrication of nanopores and show primary results of a project that is still in progress and that will be reported in further works.

ACKNOWLEDGEMENTS

This research was sponsored by the Center for Irradiation of Materials, Alabama A&M University and by the AAMURI Center for Advanced Propulsion Materials under the contract number NNM06AA12A from NASA, and by National Science Foundation under Grant N°. EPS-0447675.

REFERENCES

[1] Xiao Dengming; Li Honglei; Li Xuguang, *Proceedings of the 6th International Conference on Properties and Applications of Dielectric Materials*,1, pag. 310-313, 2000.
[2] G.M. Sessler, Introduction and Physical Principles of Electrets, Spring-Verlag, New York, 1980.
[3] Caplan M. R.; Chiang C.-Y.; Loyd D. R. and Yen L. Y., *Journal of Membrane Science*, 1997, **130**, pp. 219-237.
[4] P. Yu. Apel, Yu. E. Korchev, Z. Siwy, R. Spohr and M. Yoshida, *Nuclear Instruments and Methods in Physics Research Section B: Beam Interactions with Materials and Atoms*, **184**, pag. 337-346, 2001.
[5] Jiali Li, Derek Stein, Ciaran McMullan, Daniel Branton, Micheal J. Aziz and Jene A. Golovchenko, *Nature*, **412**, pag. 166, 2001.
[6] Derek Stein, Jiali Li and Jene A. Golovchenko, *Physical Review Letters*, **89**, pag. 1, 2006.
[7] Milner R G and Richards D 2001, *J. Microscopy*, **202**, 66, 2001.
[8] Zeisel D, Deckert V, Zenobi R and Vo-Dinh T, *Chem. Phys. Lett.*, **283**, pag. 381, 1998.
[9] A.C. Ferrari, J. Robertson, *Physical Review B*, **61**, 20 (2000).

AUTHOR INDEX

SUBJECT INDEX